Sensemaking and Neuroaesthetics

James Hutson · Piper Hutson ·
Morgan Harper-Nichols

Sensemaking and Neuroaesthetics

Neuroarts and the Spectrum of Neurodiverse Experiences

palgrave
macmillan

James Hutson
Lindenwood University
St Charles, MO, USA

Piper Hutson
Lindenwood University
St Charles, MO, USA

Morgan Harper-Nichols
Lindenwood University
St Charles, MO, USA

ISBN 978-3-031-58044-4 ISBN 978-3-031-58045-1 (eBook)
https://doi.org/10.1007/978-3-031-58045-1

Cover credit: Maram_shutterstock.com

This Palgrave Macmillan imprint is published by the registered company Springer Nature Switzerland AG
The registered company address is: Gewerbestrasse 11, 6330 Cham, Switzerland

Paper in this product is recyclable.

This manuscript is fervently dedicated to our spouses, children, families, and friends, who have been the unwavering pillars of support throughout this journey. Your infinite patience, love, and understanding have been the bedrock that sustained us through the moments of doubt and the long hours of research and writing. Your enduring encouragement and presence have not only provided sanctuary but also a profound source of inspiration, transforming this endeavor into a shared voyage of growth and discovery. Morgan Harper-Nichols extends a heartfelt thank you to her family and friends for their continuous encouragement and steadfast support in this journey. Piper and James Hutson specifically dedicate this work to their children, who have been an endless fountain of inspiration throughout the study.

We also extend our deepest gratitude to all the neurodivergent individuals who courageously shared their stories with us. Your contributions have immensely enriched this work and will undoubtedly aid in advancing self-awareness and well-being across all neurotribes. Your voices and experiences are integral to the tapestry of this research, and for this, we are eternally grateful.

Acknowledgments

We wish to express our profound gratitude to the constellation of researchers, colleagues, and professionals who have played a pivotal role in shaping this book. Their expertise and support have been invaluable in navigating the complex realms of neurodiversity, neuroarts, neuroaesthetics, and sensemaking. We are particularly indebted to Drs. Jocelyn Kiss, Colleen Biri, Sara Bagley, Stephanie Afful, and Caitlyn McGinley for their insights and unwavering dedication.

Morgan Harper-Nichols extends a special thanks to the Lindenwood University community for the invaluable mentorship and support provided throughout her academic journey. A heartfelt appreciation goes out to Dr. James Hutson, Dr. Piper Hutson, Ben Fulcher, and Andrew Smith for their guidance and significant impact on her research and academic progression.

We recognize and celebrate the tireless efforts and pioneering work of our fellow researchers in the field, whose studies have significantly informed our work and contributed to a richer understanding of the intersection between neurodiversity and artistic expression. We extend special thanks to the College of Arts and Humanities at Lindenwood University for providing the essential resources and fostering an environment conducive to interdisciplinary research, crucial in uniting the fields of neuroscience, psychology, and the arts.

Our academic peers and mentors deserve our deepest appreciation for their invaluable guidance and insightful critiques, which have sharpened our perspectives and approaches. Their wisdom has been a beacon of light throughout this endeavor. We also owe a debt of gratitude to the artists and individuals from neurodiverse communities who generously shared their experiences and creations. Their narratives and art have injected authenticity and profound depth into this exploration.

Finally, our heartfelt thanks go to the editorial team and support staff. Their meticulous attention to detail and unwavering commitment have been instrumental in refining this manuscript to its final form. This book stands as a testament to a shared vision and collaborative effort to elucidate the human experience in its rich diversity. We are truly honored to have embarked on this journey with such a dedicated and inspirational group of individuals and organizations.

Contents

About the Authors

Dr. James Hutson specializes in multidisciplinary research that encompasses artificial intelligence, neurohumanities, neurodiversity, immersive realities, and the gamification of education. Earning a Bachelor of Arts in Art from the University of Tulsa, a Master of Arts in Art History from Southern Methodist University, and a Ph.D. in Art History from the University of Maryland, College Park, he later acquired additional master's degrees in Leadership and Game Design from Lindenwood University and is presently engaged in doctoral studies in Artificial Intelligence at Capitol Technology University. Over the span of his academic career since 2006, Hutson has held various pedagogical and administrative positions across five universities, including Chair of Art History, Assistant Dean of Graduate and Online Programs, and most recently, Lead XR Disruptor and Department Head of Art History and Visual Culture. Notably, his scholarly portfolio includes several books on the application of artificial intelligence in education and cultural heritage, including *Inclusive Smart Museums* (Palgrave McMillan, 2023), as well as a plethora of articles and case studies.

Dr. Piper Hutson is a curator with over sixteen years of experience in galleries across the United States. Her career includes co-curating more than forty exhibitions, as well as serving as the head curator for ten shows over the past decade. Dr. Hutson's expertise lies in nineteenth-century British art, art education, and inclusivity in cultural heritage collections for neurodiverse populations and she currently serves as both a Corporate Art Curator for Wells Fargo and an instructor in higher education at Lindenwood University. Her current research is dedicated to improving inclusivity in cultural heritage collections for neurodiverse populations. She has authored several works on inclusivity in the workplace and best practices for supporting neurodiversity in cultural heritage institutions with her research underscoring the importance of creating accessible and welcoming spaces for all individuals to engage with cultural heritage collections. Most recently, her studies on neurodiversity and arts include *Inclusive Smart Museums* (Palgrave McMillan, 2023).

Morgan Harper-Nichols is an Atlanta, GA-based neurodivergent, synesthesia artist (autistic/ADHD/sensory processing disorder) with a client roster that includes Google, Meta, Apple, Adobe, Starbucks, and more. She is currently pursuing an MFA in Interdisciplinary Media Arts at Lindenwood University, where she concentrates on interactive art and games. She is also focused on studying the intersection of neurodiversity, art, its impact at a sensory level, and artificial intelligence. With a creative career spanning over a decade, her areas of expertise include digital art, graphic design, interactive platforms, AI, 3D art, and virtual environments. She is the creator of the "Storyteller" app, available on iPhone and Android. Beyond her creative work, Morgan serves as the Vice President of the Board of Directors for the mental health organization TWLOHA and has been a featured speaker at various events, including the 2023 United Nations World Autism Awareness Day and Google Talks. Committed to mental well-being, accessibility, and a deeper understanding of neurodiversity in the arts, her aim is to create engaging, sensory-inclusive interactive experiences and contribute to interdisciplinary media arts education.

List of Figures

List of Tables

PART I

Foundations

CHAPTER 1

Introduction: The Nexus of Neuroarts, Neuroscience, and Technology

The introduction lays the foundation for understanding how neuroarts—a fusion of neuroscience and art, enhanced by technology—is reshaping creativity. Tracing neuroarts evolution from a novel idea to a significant interdisciplinary field, this chapter delves into the synergy between the brain sciences and artistic expression and how technology can act as a bridge, expanding the boundaries of how we create and perceive art. Central to this discussion is the concept of sensemaking, which involves interpreting and making sense of various sensory experiences. A key focus is on the diverse sensory experiences within neurodiversity, acknowledging how different neurological experiences can influence art. The concept of neurotribes is introduced to stress the importance of recognizing diverse cognitive profiles, especially in the context of neurodiversity, art, and human connection. The chapter explores current research in neuroaesthetics, particularly in relevance to Autism Spectrum Conditions (ASC), shedding light on strength-based approaches to understanding unique artistic perceptions within the Neurodiversity umbrella. It concludes by discussing theoretical frameworks like phenomenology and enactivism, and introduces synesthesia—a condition where sensory experiences are interconnected resulting in a unique sensory perspective and prevalence of individuals preference through an evolution of sensory experiences.

J. Hutson et al., *Sensemaking and Neuroaesthetics*,
https://doi.org/10.1007/978-3-031-58045-1_1

1.1 The Brain on Art

Research continues to reveal the profound effects of art on the human brain, particularly in the context of well-being (Cotter et al., 2023; Mastendrea et al., 2019; Secker et al., 2018). The burgeoning fields of neuroarts and neuroaesthetics, situated at the intersection of neuroscience, art, and technology, are predicated on this very relationship (Magsamen & Ross, 2023). For instance, the act of engaging with art, especially colorful paintings (Fig. 1.1), has been found to significantly enhance mental wellness in certain cases, underscoring the idea that the aesthetic appreciation of art is not merely a passive experience but an active process involving complex neurological pathways (Puranik, 2023). Colorful artwork, whether experienced in person or online, is found to trigger the release of dopamine, a neurotransmitter associated with pleasure and reward. This release leads to feelings of happiness and well-being, akin to the sensations evoked by other pleasurable experiences such as food or sex. These findings suggest a universal response to aesthetic stimuli, underscoring the essential role of art in human life.

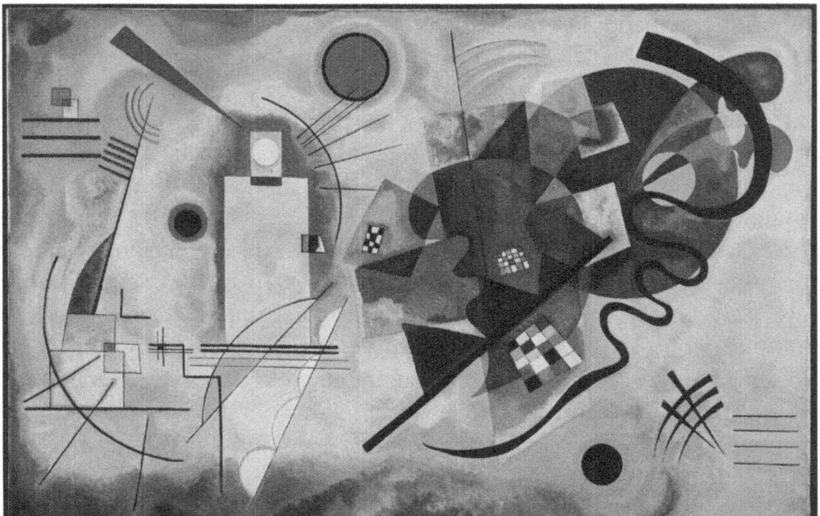

Fig. 1.1 Wassily Kandinsky, *Gelb, Rot, Blau*, 1925 (*Source* From Wikimedia Commons, licensed under CC0)

The scientific exploration of how the brain responds to art has garnered increasing interest, particularly because the disciplines of art and brain science have historically developed independently. Art history and criticism, grounded in the humanities, have traditionally concentrated on the aesthetic evaluation, cultural importance, and interpretation of art. In contrast, neuroscience has aimed to comprehend the functions and structures of the brain using scientific methods. Zeki (1999, 2007) represents a previous attempt at understanding such reactions, and their research reveals that viewing beautiful paintings can be equivalent to looking at someone we love, activating the same areas of the brain responsible for pleasure sensations. This phenomenon can be attributed to the engagement of brain structures like the orbitofrontal cortex, amygdala, and ventral striatum (Fig. 1.2) when encountering aesthetically pleasing art. At the same time, the influence of art extends beyond mere visual pleasure, contributing significantly to mental and emotional well-being. Engaging with art, particularly vibrant and colorful pieces, has been shown to reduce stress and promote relaxation (Raad et al., 2021; Smyth et al., 2020). Additionally, it can enhance cognitive functions, notably in areas such as memory retention and critical thinking. The practice of art can cultivate mindfulness and aid in sharpening concentration. It is advocated that regular engagement with artistic experiences should be an integral part of a comprehensive approach to well-being. This deep, mindful interaction with art is not only soothing but also instrumental in heightening emotional responses and improving mental health (Windle et al., 2018).

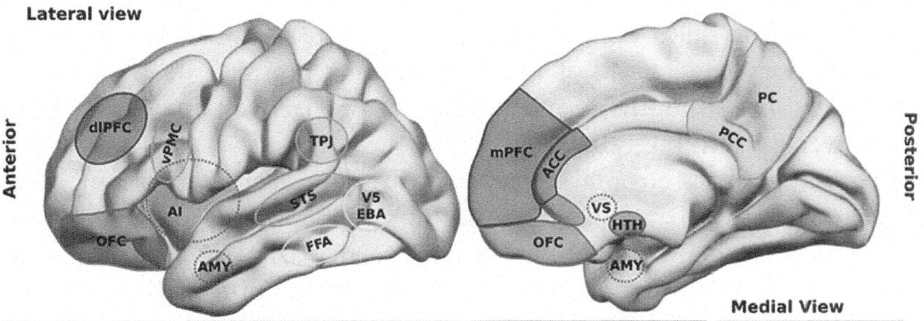

Fig. 1.2 Billeke and Aboitz, brain areas that participate in social processing 2013 (*Source* From Wikimedia Commons, licensed under CC-3)

Furthermore, art has the capacity to activate the prefrontal cortex, a critical region of the brain involved in focus and interpreting meaning from sensory information. This aspect of the impact art has on the brain underscores its role in cognitive processing and emotional regulation (Bigliassi, 2022). The capacity of art to stimulate the prefrontal cortex—a key area in the brain responsible for focus and deciphering meaning from sensory data—holds particular significance for individuals who are neurodivergent. While processing of sensory stimuli has a range of neurodiverse experiences, these individuals are prone to specific conditions. Neurodiversity as a term reflects the understanding that variations in the neurocognitive functioning of the human brain are a natural and valuable aspect of human diversity (Dwyer, 2022). This concept encompasses a range of neurological differences, viewing them not as deficits, but as normal variations in the human genome. In this context, the term "neurotypical" refers to individuals whose neurocognitive development and functioning are consistent with the societal standard. These individuals typically do not exhibit significant deviations in their neurological processing (Radulski, 2022). In contrast, "neurodivergent" describes those whose neurological development and functioning diverge from this conventional standard. This divergence is not a mark of inferiority, but rather a reflection of the diverse spectrum of human brain functioning. The neurodivergent group includes individuals with various conditions, such as Autism Spectrum Condition (ASC), characterized by distinct social interaction, communication styles, and sensory sensitivities; Attention-Deficit/Hyperactivity Disorder (ADHD), noted for its unique patterns of attention, hyperactivity, and impulsivity; Dyslexia, which primarily impacts reading and language processing; Dyspraxia, affecting motor skill development; and Tourette Syndrome, known for its involuntary tics. Each of these conditions presents its own set of challenges and strengths, contributing to the rich tapestry of human neurodiversity (Hulbig, 2023).

Regarding the prefrontal cortex, which plays a crucial role in executive functions such as planning, decision-making, and social behavior regulation, engagement with art can be particularly beneficial for individuals with neurodivergent conditions. This interaction with art has the potential to enhance these executive functions, offering therapeutic and supportive advantages. To exemplify this, the neuroarts installation in Milan by Google hardware design group, A Space for Being (2019), serves as a pertinent example. The exhibit demonstrates how different sensory environments elicit varied physiological responses, highlighting the significance of being attuned to one's physiological and aesthetic surroundings (Uszerowicz, 2019). In other words, the different installations allow visitors to measure which interiors are best suited for their health and well-being by gaining insights into how their bodies and minds react to different artistic and sensory experiences, emphasizing the importance of understanding and harnessing these responses for enhanced well-being and cognitive engagement. For example, people with autism might process the emotional aspects of art in distinctive ways, which could help them develop a better understanding of emotions and empathy. Likewise, individuals with ADHD may find that the concentrated involvement needed for creating or appreciating art offers a well-organized, yet imaginative, outlet for their energy and thoughts.

Additionally, the exploration of neurodiversity and sensory experiences within the context of neuroarts inherently aligns with the social model of disability. This model posits that disability arises not inherently from an individual's physical or neurological conditions but from the societal barriers and lack of accommodations for these differences (Oliver, 1996). The social model serves as a foundational framework for understanding how neurodiverse conditions, particularly autism, are conceptualized within the broader discourse of inclusion and accessibility in the arts. However, it is also crucial to acknowledge that the social model of disability, while transformative in promoting a shift from medical pathology to social inclusivity, has been the subject of debate. Critics argue that it may oversimplify the complex interplay between biological, psychological, and social factors that contribute to the lived experience of disability (Shakespeare, 2014). Moreover, the model's emphasis on societal change, while vital, does not fully encapsulate the personal narratives and self-perceptions of individuals with disabilities, which can vary widely and encompass a range of attitudes toward their own neurodivergent traits or disabilities (Goodley, 2016). To provide a balanced perspective, it's important to integrate the biopsychosocial model, which recognizes the importance of biological conditions and personal experiences in conjunction with social barriers (Engel, 1977). This model offers a more nuanced understanding of disability, acknowledging that individual experiences of disability are shaped by a combination of genetic, psychological, and social factors.

Furthermore, engagement with the concept of neurodiversity itself requires a critical examination of how diverse neurological conditions are valued and represented in society. The neurodiversity movement advocates for the acceptance and celebration of neurological differences as natural variations within the human genome, challenging the notion of neurotypicality as the standard (Singer, 1999). Yet, this perspective too invites critical discourse on how best to support individuals with neurodivergent conditions in a manner that respects their autonomy and individual needs while striving for societal accommodation and inclusion. Incorporating these diverse perspectives enriches our understanding of the complex landscape of neurodiversity and disability. It invites ongoing dialogue and research into the ways neuroarts can serve as a medium for expressing, understanding, and navigating the multifaceted experiences of neurodivergent individuals. By embracing a multiplicity of models and viewpoints, the field of neuroarts can contribute to a more inclusive and empathetic society that values and accommodates a spectrum of sensory experiences and neurological conditions.

1.2 Art and Health

While the modern field of neuroarts and the neuroscience underlying the connections between art and well-being can be traced to the late twentieth century, the practices of using art for healing and well-being extend back to antiquity (De'Souza, 2012). In ancient cultures, the therapeutic and transformative power of art was widely recognized. For example, art was deeply embedded in religious and healing practices in ancient Egyptian culture. The Egyptians revered the healing power of visual aesthetics, evident in their elaborate tomb paintings (Fig. 1.3) and sculptures that were believed to ensure a peaceful afterlife and provide protective benefits (Allen, 2005). Similarly, in ancient Rome, healing temples, known as Asclepieia, were richly adorned with art, integrating music, theater, and visual arts as essential components of patient healing regimens (Fig. 1.4) (Burge, 2015). The Greek philosopher Aristotle acknowledged the cathartic impact of drama, particularly tragedy, advocating its emotional purging and healing virtues (Ferrari, 2019). During the Renaissance, a deeper exploration of the relationship between art and health emerged, significantly influenced by the humanist movement. This period highlighted the importance of individuality and holistic well-being with "outward and inward senses" given equal weight, and viewing art as a symbol of balance, harmony, and beauty essential for uplifting the spirit and enhancing mental well-being (Fig. 1.5). Renaissance art,

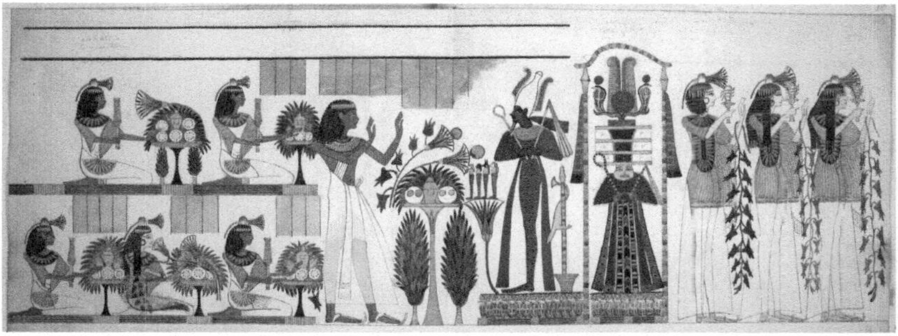

Fig. 1.3 Egyptian tomb wall-painting—Egyptian collections, vol. XI (1826–1838) (*Source* From Wikimedia Commons, licensed under CC0)

with its detailed portrayal of human emotions and the natural world, was considered a channel for emotional and spiritual enrichment (Thibault, 2019).

Moreover, many indigenous cultures have longstanding traditions of integrating art into healing practices. Native American cultures, for instance, use sand painting in healing ceremonies, where the intricate and symbolic designs are believed to restore balance and harmony (Samuels, 1995). In ancient Chinese culture, the practice of calligraphy (Fig. 1.6) was seen as a meditative and healing exercise, with the careful strokes believed to harmonize one's qi and promote mental tranquility (Zhang & Rose, 2001). On the Indian subcontinent, the traditional art of creating Rangoli (Fig. 1.7), involving intricate floor patterns made with colored powders or rice, was a part of religious and healing rituals, believed to bring good fortune and health (Surapaneni, 2023). Additionally, Aboriginal Australian art, known for its intricate dot paintings (Fig. 1.8) depicting sacred stories and rituals, has been used for thousands of years as a tool for spiritual healing and connection to the land and ancestry (Bennett, 2023). This historical perspective illustrates that the intersection of art and well-being, now increasingly illuminated through the lens of neuroscience, is deeply rooted in centuries of human culture, philosophy, and diverse global traditions.

The advancement of cognitive psychology and neuroscience has established new frameworks for comprehending the brain's information processing mechanisms, encompassing both perceptual and emotional responses. This significant shift in perspective has enabled a more detailed

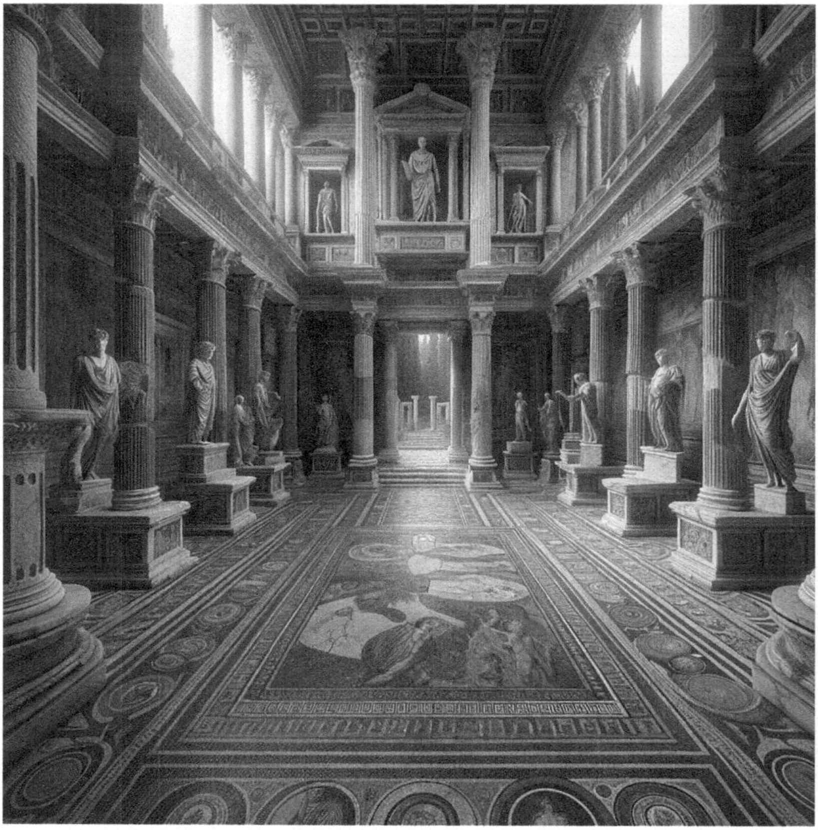

Fig. 1.4 Recreation of Ancient Roman Aesclepieia, 2023 (*Source* Permission by authors)

investigation into the brain's perception and interpretation of art, effectively merging psychological theories with neurological findings. With the experience of visual art being associated with health benefits firmly rooted in culture, there is little surprise that it is now being prescribed by medical professionals and psychologists to treat a range of ailments. The contemporary concept of "Arts on Prescription," described by Golden (2023) represents a continuation of this tradition of leveraging the healing power of art, though innovative for mental healthcare. In the 1940s, one of the first acknowledgments of art's therapeutic benefits in a medical

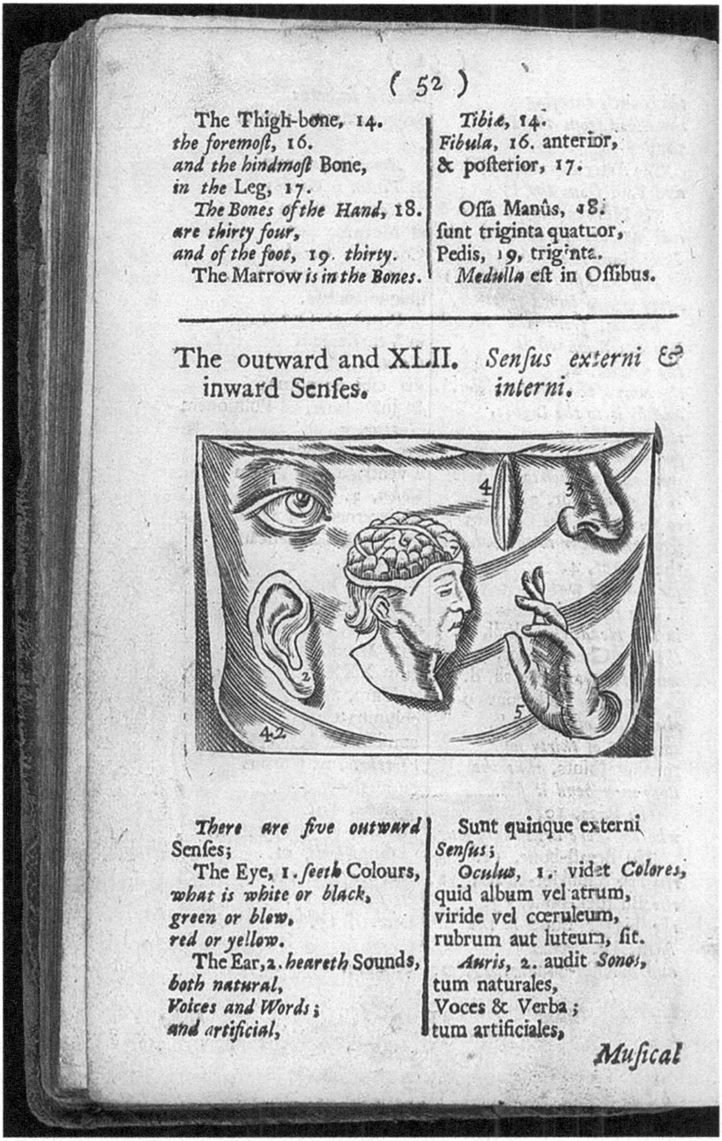

(52)

The Thigh-bone, 14. *the foremost*, 16. *and the hindmost* Bone, *in the* Leg, 17.	*Tibia*, 14. *Fibula*, 16. anterior, & posterior, 17.
The Bones of the Hand, 18. *are thirty four*, *and of the foot*, 19. *thirty*.	Ossa Manûs, 18. sunt triginta quatuor, Pedis, 19, triginta.
The Marrow *is in the Bones*.	*Medulla* est in Ossibus.

The outward and inward Senses. XLII. *Sensus externi & interni*.

There are five outward Senses;	Sunt quinque externi *Sensus*;
The Eye, 1. *seeth* Colours, *what is white or black*, *green or blew*, *red or yellow*.	*Oculus*, 1. videt *Colores*, quid album vel atrum, viride vel cœruleum, rubrum aut luteum, sit.
The Ear, 2. *heareth* Sounds, *both natural*, *Voices and Words*; *and artificial*,	*Auris*, 2. audit *Sonos*, tum naturales, Voces & Verba; tum artificiales,

Musical

Fig. 1.5 The outward and inward senses. John Amos Comenius, *Visible world: or, a nomenclature, and pictures, of all the chief things that are in the world, and of men's employments therein*. 1728 edition (*Source* From Wikimedia Commons, licensed under CC-BY 4.0)

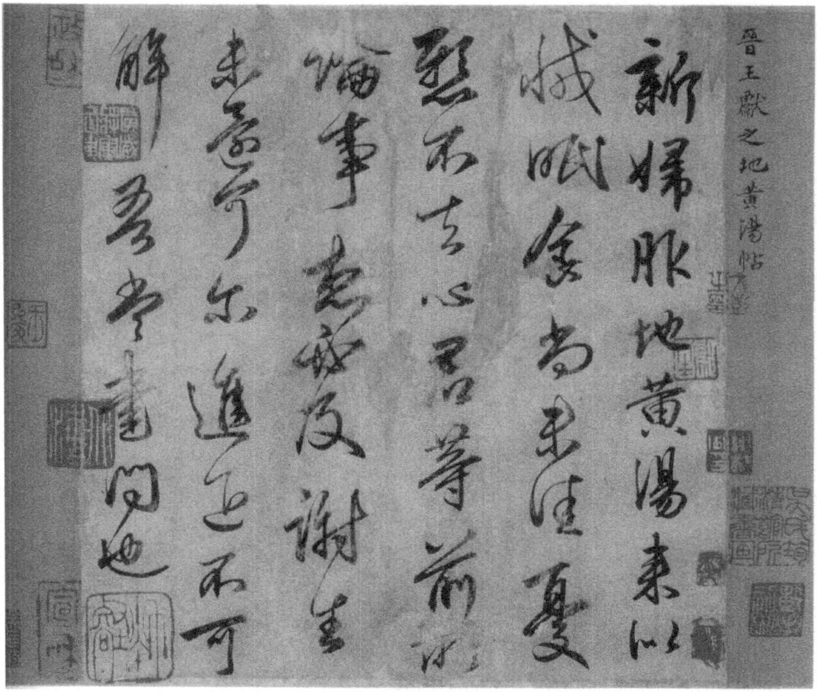

Fig. 1.6 Wang Xianzi imitation by Tang Dynasty (*Source* From Wikimedia Commons, licensed under CC0)

context is linked to Dr. Adrian Hill (1895–1977), a British artist. While recuperating from tuberculosis, Hill experienced firsthand the therapeutic effects of art and began integrating it into patient care. He introduced the term "art therapy" in 1942, marking a significant step in recognizing art's healing potential, particularly in hospitals. His work laid the foundation for establishing art therapy as a formal field (Chapin & Fish, 2016). Simultaneously in the United States, Margaret Naumburg (1890–1983), often hailed as the "mother of art therapy," championed the use of art as a psychotherapeutic expression tool. She underscored the value of art in articulating the unconscious mind, employing it therapeutically with children and adolescents in psychiatric settings. Her advocacy and application of art in psychotherapy contributed significantly to the development and recognition of its therapeutic benefits (Junge, 2015).

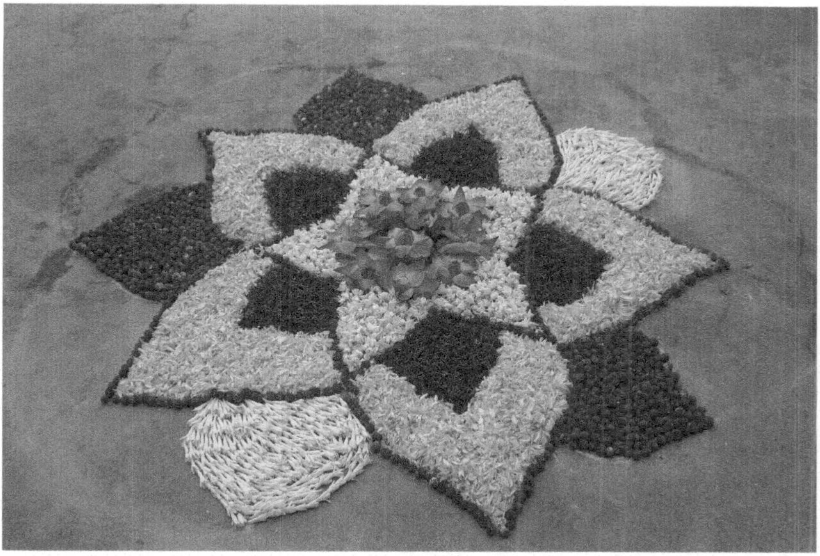

Fig. 1.7 Flower Rangoli Chennai (*Source* From Wikimedia Commons, licensed under CC0)

The model of social prescribing, integral to this approach, allows healthcare providers to refer patients to local arts and cultural resources, alongside traditional treatments. By addressing the broader social drivers of health, this model expands the conventional healthcare paradigm. The concept, having roots in the United Kingdom, has evolved to encompass a variety of services, including arts, which have shown significant benefits in reducing anxiety and depression, and improving the quality of life. In the United States, initiatives like the CultureRx: Social Prescription Pilot in Massachusetts demonstrate the practical application of the model, with healthcare professionals prescribing cultural activities as part of patient care (Golden et al., 2023). Arts on prescription not only broadens the scope of resources for healthcare providers but also emphasizes the necessity for equitable access to arts and cultural activities. Providing a range of artistic mediums accommodates different interests and skill sets. For instance, those facing challenges with fine motor skills might find activities like painting or sculpting more suitable, whereas individuals with dyslexia may lean toward visual arts rather than literary forms. Additionally, for

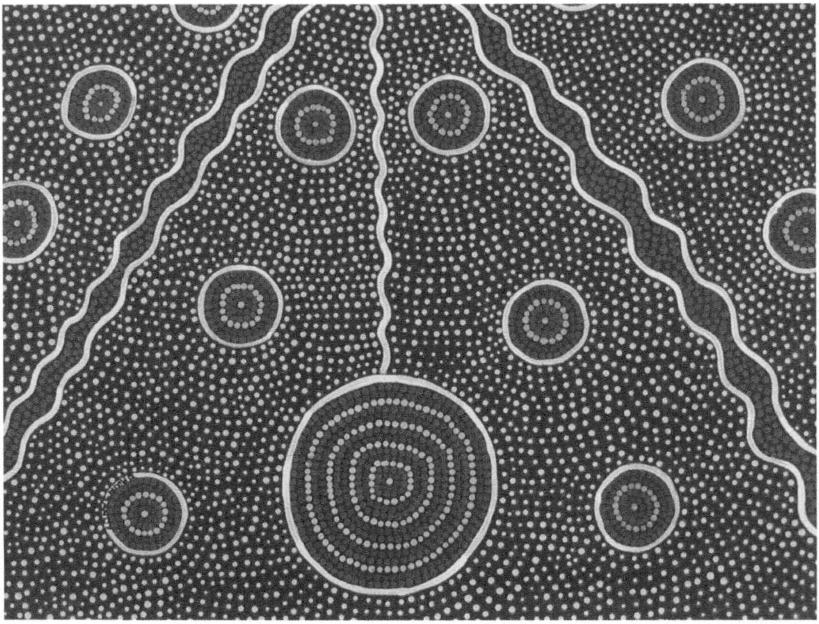

Fig. 1.8 Aboriginal painting, 2014 (*Source* From Wikimedia Commons, licensed under CC0)

neurodivergent individuals who may struggle with social communication, the collaborative aspect of creating art together offers an alternative way to engage and connect with others.

1.3 Complexities of Sensory Experience

It should be pointed out, however, that examples thus far have predominantly focused on visual artistic experiences, the field of neuroarts transcends this singular sensory perspective. A key factor challenging the traditional visual dominance is the concept of *ocularcentrism*, which has been critically examined in the context of sensory design and phenomenology. Coined by Finnish architect Juhani Pallasmaa (2012), ocularcentrism refers to the predominance of visual form and the Western preoccupation with sight as the primary means of experiencing the world. While this visual bias has long been the cornerstone of modern science and perception, research has continued to investigate the various intersections, overlaps, and crossovers between the five senses of humans—sight, hearing, taste, smell, and touch. Art frequently mirrors the norms, values, and issues prevalent in different cultures and societies. By engaging with art, people can gain a deeper understanding of various cultures and historical eras, thereby increasing their cultural awareness and empathy. Furthermore, art has the power to question societal norms and stimulate conversations about social issues and inherent biases, consequently shaping the way individuals perceive and interact with their society. Examples such as those discussed by Colino (2023) demonstrate that senses can extend into one another, such as the visual into the auditory realm. As Colino outlines, colored noises, including white, pink, brown, green, blue, and purple, each bears unique characteristics and exert various effects on human well-being. White noise, encompassing all frequencies audible to humans, aids in sleep improvement and focus. Pink noise, with its emphasis on lower frequencies, enhances sleep quality and memory retention. The low tones of brown noises aid in concentration, while green noise, often likened to natural sounds, has a calming effect. Conversely, blue and purple noises, characterized by higher frequencies, are generally not recommended for relaxation due to their more disruptive nature. These auditory experiences demonstrate the intricate and diverse ways in which neuroarts can engage with sensory modalities beyond visual stimuli.

Participating in creating or appreciating art significantly contributes to the formation of personal identity and self-awareness. This engagement provides an avenue for expressing and exploring oneself, enabling individuals to delve into different facets of their personality, experiences, and beliefs. For numerous people, art is integral to their identity and serves as a vital channel to convey their distinct viewpoints to the world. With such potential for various forms of sensory modalities to be experienced, more attention is being paid to unexpected sensory stimuli. The potential is made possible through the conscious use by the mind of external sensory stimuli. Recent research from cognitive neuroscience asserts that neurotypical individuals (those not considered neurodivergent) consciously experience only a fraction of their mental activities in such a way, which opens the door to a deeper understanding of the unconscious realms of sensory experiences (Lupton & Lipps, 2018). This insight is crucial to the field of neuroarts, which delves into these unseen mental activities and their significant impact on perception, cognition, and overall human experience. But before continuing, a deeper understanding of the field is warranted.

1.4 Neuroarts and Neuroaesthetics

Neuroarts, a rapidly expanding transdisciplinary field, represents the confluence of the arts and scientific insights into their effects on the body, brain, and behavior (Siler, 2015). This innovative domain emphasizes how engagement in artistic activities can lead to tangible improvements in health and well-being. Encompassing a diverse array of art forms—ranging from music and dance to poetry and architecture—neuroarts explores the unique interactions each art form has with the brain and body (Magsamen & Ross, 2023). The World Health Organization (WHO) has recognized the significant contributions of the arts to the prevention and treatment of both mental and physical health conditions (Fancourt & Finn, 2019). As neuroarts continues to evolve, there is a growing imperative for comprehensive research to optimize art therapies, determining their most effective delivery, appropriate duration, and customization to individual needs.

As noted, art has historically been integral to health, well-being, and community cohesion across various cultures, particularly during times of stress, showcasing an innate human connection to artistic expression (Macnaughton et al., 2005). The connection is now being reinforced by scientific research. Technological advancements, including brain imaging, portable devices, and wearable sensors, have empowered researchers to observe and quantify the response of the brain to artistic stimuli in real time (Cheung et al., 2019). In turn, the research is enhancing our understanding of the profound impact of art. Neuroarts illuminates the involvement of multiple brain systems when engaging with art, activating reward pathways, motor functions, sensory processing, and perception, which, in turn, capitalizes on the capacity of the brain for neural plasticity and its extensive neural networks (Von Bernhardi et al., 2017).

As a result, art is increasingly acknowledged as a complementary element in traditional medical practices, such as Arts on Prescription, with growing evidence of its efficacy in enhancing mobility, memory, speech, mental health, resilience, and disease prevention (Garzón Chiriví, 2015). As will be discussed, various art forms have demonstrated specific health benefits; for instance, music therapy can improve cognitive functions in dementia patients (Zhang et al., 2017), dance therapy can alleviate symptoms in Parkinson's disease (Michels et al., 2018), and poetry can offer solace in end-of-life care (Clary, 2010). Even architecture contributes to creating healing environments (Singh et al., 2021).

On the other hand, neuroaesthetics, a closely aligned field, delves into understanding the neural mechanisms behind our aesthetic experiences with art. It explores how different art forms stimulate emotional and cognitive responses, providing a scientific basis for the impact observed in neuroarts (Chatterjee & Vartanian, 2014). Together, neuroarts and neuroaesthetics form a dynamic, interdisciplinary arena that not only deepens our scientific understanding of the brain's response to art but also enhances practical applications in art-based therapies and educational initiatives. This synergy underscores the invaluable role of art in both our personal lives and broader societal health contexts. The current terrain of neuroaesthetic inquiry includes examining the design of environments, such as the Israel Cancer Hospital, which emphasizes the healing power of sensory-rich surroundings (Magsamen & Ross, 2023). Neuroaesthetics also explores conditions like ASC, delving into the neurophysiological underpinnings of aesthetic experience in individuals with autism. This involves a detailed examination of the structure of the cerebrum, encompassing the frontal, temporal, parietal, and occipital lobes, which are responsible for executive functioning, memory formation, and sensory processing (Mather, 2013). Additionally, the role of the cerebellum in balance, movement, coordination, and procedural memory is considered, alongside the connection to the brainstem and the autonomic nervous system, including its sympathetic and parietal divisions that govern reactions like fight or flight and rest and reset.

The field further discusses the role of emotions in our responses to environmental stimuli, as outlined by neuroanatomist Taylor (2021). This includes the psychological and physiological impacts of stress and the body's response to prolonged stress. Neuroaesthetics also explores sensory differences and their associations, such as the pleasant sensation of ASMR (autonomous sensory meridian response) and the discomfort of sensory overload. The senses play a crucial role in survival, acting as protective mechanisms by alerting us to potential danger (Del Campo & Kehle, 2016). Other organizations have sought to institutionalize the study of the field. For instance, The NeuroArts Blueprint initiative, launched in partnership between the Johns Hopkins International Arts + Mind Lab and the Aspen Institute's Health, Medicine & Society Program, exemplifies the advancements in this field. Guided by leaders across various disciplines, the initiative aims to cultivate an ecosystem for neuroarts. This involves creating research and practice pathways, securing funding, implementing supportive policies, and building infrastructure capacity. The ultimate goal is to bring together science, arts, and technology to improve health and well-being globally (Magsamen et al., 2023).

When considering the role of the visual cortex in neuroaesthetics, it becomes apparent that this brain region does more than process visual information from the eyes. It interprets what we see, connecting it with our memories and emotions. People with more art experience exhibit more complex neural responses to art, indicating a deeper appreciation and understanding. The visual cortex is divided into specialized sections, such as V4, which processes color. This suggests that variations in this brain region, such as in colorblind individuals, could affect how art is experienced (Roe et al., 2012). Thus, neuroaesthetics, or neuroarts, underscores the transformative power of art on the brain. It demonstrates that art engagement goes beyond aesthetic appreciation, actively exercising the brain and contributing to neuroplasticity. This process can lead to increased intelligence, empathy, and emotional intelligence, making art an essential component of a well-rounded human experience. As research continues, the differences in brain responses to virtual and real-world stimuli, particularly in areas like the hippocampus and prefrontal cortex, are also being explored, providing a deeper understanding of our interaction with art and environments.

1.5 Introduction to Synesthesia

Explorations of the intersections of neuroarts and neuroaesthetics, where art meets the intricate workings of the brain, naturally lead to another phenomenon of crosspollination—*synesthesia*. This condition, residing at the crossroads of sensory perception and cognitive experience, exemplifies the profound and often unexpected ways in which our brains can interpret and intertwine sensory information. Synesthesia, a condition where the stimulation of one sensory or cognitive pathway leads to automatic, involuntary experiences in a second sensory or cognitive pathway, presents a vivid illustration of the potential of the mind for multisensory integration (Ward, 2013). These sensory inputs are not isolated, such as only hearing sound or feeling touch, but interconnected in unique and often unexpected ways.

The developmental trajectory of multisensory functioning presents a rich area of study that intersects with various aspects of cognitive, neurological, and sensory development. A comprehensive understanding of this trajectory is essential for grasping how sensory experiences and neurodiverse conditions evolve and interact over the lifespan. From infancy through adulthood, individuals experience significant changes in how they process and integrate sensory information. The development of multisensory integration (MSI) capabilities, which refer to the brain's ability to combine information from different sensory modalities into a cohesive perceptual experience, is crucial for interpreting and navigating the world effectively (Bahrick & Lickliter, 2014). Research indicates that the foundations for MSI are present early in life but continue to mature through adolescence, influenced by both genetic factors and environmental interactions (Lewkowicz, 2014). Developmental disorders, including autism and sensory processing disorders (SPD), highlight the variability in sensory developmental trajectories. For instance, children with ASD often exhibit atypical MSI, which can affect their social communication and interaction skills (Marco et al., 2011). Similarly, individuals with SPD may have difficulties processing and responding to sensory stimuli in a manner that is considered typical, impacting their daily functioning and well-being (Miller et al., 2007).

The developmental perspective emphasizes the importance of early detection and intervention. Tailored therapeutic approaches, including sensory integration therapy and multisensory environments, have been shown to support sensory development and improve outcomes for individuals with neurodiverse conditions (Case-Smith & Arbesman, 2008). Moreover, understanding the developmental nuances of sensory processing can inform the design of inclusive educational strategies and support systems that accommodate a wider range of sensory preferences and needs. Future research should continue to explore the developmental aspects of multisensory functioning, with an emphasis on longitudinal studies that can provide deeper insights into how these processes unfold over time. Such studies are critical for developing more effective interventions and supports that can adapt to the changing needs of neurodiverse individuals throughout their lifespan.

The neural underpinnings of synesthesia suggest an enhanced level of connectivity and communication between different brain regions, explaining the extraordinary sensory experiences of *synesthetes*. A synesthete is an individual who experiences synesthesia, a neurological condition characterized by the blending of sensory or cognitive pathways (Glasser, 2023). In such multisensory integration, stimulation in one sensory or cognitive domain, such as hearing music, involuntarily and automatically triggers a perception in another, like seeing colors. This sensory crossover allows synesthetes to experience the world uniquely, with sounds being perceived as colors, letters, or numbers imbued with inherent colors, and even tastes taking on shapes. For example, those with *chromesthesia* experience a fusion of sound and color, where specific musical notes or sounds evoke distinct colors, transforming melodies into visual spectacles (Nerurkar et al., 2022). Another form, *grapheme-color synesthesia*, involves perceiving letters or numbers as colored, a phenomenon tied to the role of the fusiform gyrus in color processing and character recognition (Hoffman et al., 2019). Additionally, *mirror-touch synesthesia* expands the understanding of sensory connections by enabling individuals to physically feel the sensations others experience, linking to the empathy circuits of the brain in areas like the premotor cortex and superior temporal sulcus (Ward, 2019). This heightened neural interconnectivity challenges traditional views of sensory processing and perception, offering a window into a more integrated and interconnected sensory world.

The history of synesthesia in popular culture, fine art, and medical fields is as rich and varied as the condition itself, intertwining artistic expression, cultural representation, and scientific inquiry. In the realm of fine art, Wassily Kandinsky (1866–1944), a pioneer of abstract art (Fig. 1.1) himself had four coupled senses. Through these perceptual convergences, Kandinsky made significant contributions with his long essay *Concerning the Spiritual in Art* (1914, 2012) where he attempted to devise a universal algorithm for translating sensory experiences across different modalities. While he was ultimately unsuccessful in his efforts, the insights highlighted that synesthetic pairings are, in fact, highly individualistic and idiosyncratic, rather than universal, a realization underscored later in the writings of neurologist Richard Cytowic (1952–) (2018).

Popular culture has also embraced synesthetic concepts. Walt Disney's *Fantasia* (1940) is a classic example, built on the idea of sound-to-sight correspondences, visually representing music in imaginative and colorful animations (Disney's, W., 2023). Similarly, *The Wizard of Oz* (1939), produced by Metro-Goldwyn-Mayer, utilized color tinting effects, notably in the "horse of a different color" scene, to achieve a form of perceptual synesthesia, seamlessly blending auditory and visual elements (Schlich, 2012). The phenomenon has been observed and remarked upon throughout history, even outside the scientific community. Ellen Emerson (1839–1909), the daughter of American philosopher Ralph Waldo Emerson (1803–1882), spoke of colored words when she was just eight years old, indicating an early awareness of synesthetic experiences (Jewanski et al., 2013).

As with Ellen, engagement of the medical field with synesthesia dates back to the early nineteenth century. The first reported clinical case was documented in 1812 in a Latin medical dissertation by George Tobias Ludwig Sachs (1786–1814), a polymodal synesthete who experienced colors for letters, musical tones, numbers, and days of the week (Jewanski et al., 2011). Later, in 1880, polymath Sir Francis Galton (1822–1911), cousin of evolutionary biologist Charles Darwin (1809–1882), wrote about visualized numerals in the journal *Nature* (Greenberg, 2009). Significantly, he noted the familial occurrence of synesthesia, suggesting a genetic component. The first German book on the subject, *Das Farbenhören und der synästhetische Faktor der Wahrnehmung* (1927) by Annelies Argender, further explored this fascinating condition. Interestingly, the science of behaviorism, dominant between 1920 and 1940, temporarily overshadowed the study of synesthesia, focusing more on observable behaviors rather than introspective experiences. This period marked a significant shift in the approach to understanding synesthesia, moving away from the nineteenth-century practice of distinguishing symptoms from signs (Birnbaum, 1955). As noted, it would not be until the 1990s that the conscious introspection of experience regained prominence over the mere observation of behavior, allowing for a deeper understanding and appreciation of synesthesia (Cytowic, 1995). The shift marked a turning point in the scientific and cultural exploration of cognitive-sensory intersections, reaffirming its significance as a unique and enlightening window into the human sensory experience. Still, critics noted the lack of evidence and would deride those who published on the topic (Van Campen, 2010).

By the early 2000s, the long-awaited brain imaging evidence critical to understanding synesthesia was finally available, leading to significant shifts in perception within the scientific community (Grossenbacher & Lovelace, 2001; Rouw & Scholte, 2007; Whitaker et al., 2014). The newly available abundance of brain imaging data not only silenced critics but also overturned longstanding theories about brain organization (Fig. 1.9). The paradigm shift underscored the importance of synesthesia, elevating it from a mere curiosity to a crucial window into the vast complexities of the mind and brain. This revelation paved the way for investigating the neural underpinnings of interlinked sensations and their potential connections with conditions like autism, which also involve unique sensory processing patterns (Riedel et al., 2020). Synesthetes typically experience a blending of senses, whereas individuals with autism may display heightened or diminished reactions to sensory stimuli. This similarity points to a possible neural connection at the level of sensory processing (Neufeld et al., 2013). Neuroimaging studies have particularly highlighted parallels in brain activation patterns in areas related to sensory processing in both synesthetes and autistic individuals, deepening our understanding of these conditions (Bouvet et al., 2014, 2019; Van Leeuwen et al., 2021). Furthermore, the variations and forms of synesthesia are diverse with neuro-sensory interplay being bimodal, involving two sensory areas, or multimodal, affecting multiple senses. Examples include chromesthesia, where sounds trigger color perceptions, and grapheme-color synesthesia, where letters or numbers are inherently colored. More complex forms involve tactile sensations triggered by sounds or sights, and audiomotor synesthesia, where sounds induce bodily movements (Nanay, 2020).

At this point, a major distinction should be noted in synesthesia is between sensory and cognitive forms. Sensory synesthesia involves direct sensory triggers, such as colors evoked by sounds, while cognitive synesthesia combines sensory experiences with semantic triggers like letters or numbers, producing colored perceptions of linguistic elements (Hubbard, 2007). Conceptual synesthesia extends to abstract concepts being perceived as colors or shapes. Recent advancements in neuroimaging and other physiological tools, like tensor diffusion imaging (Fig. 1.10) and magnetoencephalography, have allowed for deeper insights into the neural networks and cross-connectivity underlying synesthesia (Nowak & Gimenez, 2015). These tools help understand how neural networks establish, self-calibrate, and disband, offering a window into the dynamic processing capabilities of the brain.

1.6 Finding Your Neurotribe

The advances made in the early 2000s in brain imaging assisted in understanding synesthesia and have had a profound impact on the development of digital interventions tailored to various neurotypes. This progress

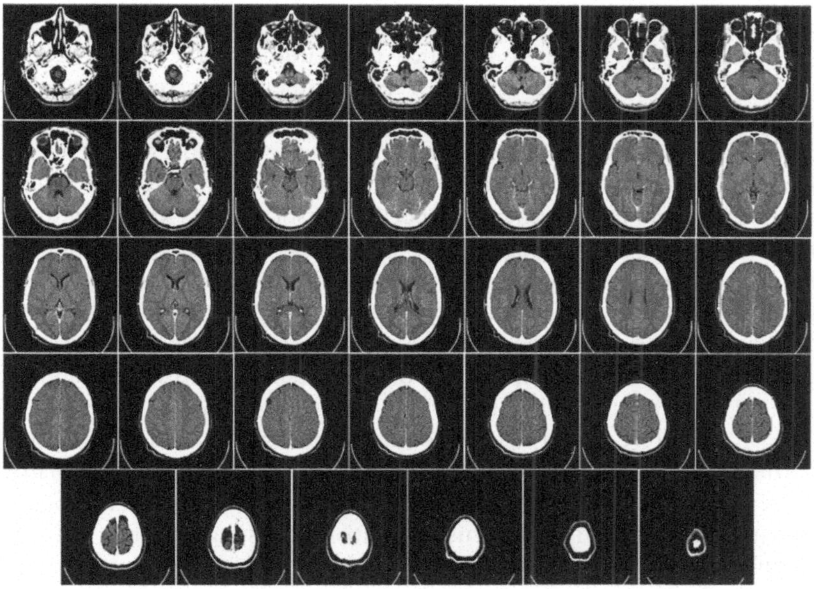

Fig. 1.9 Computed tomography of the human brain (*Source* From Wikimedia Commons, licensed under CC0)

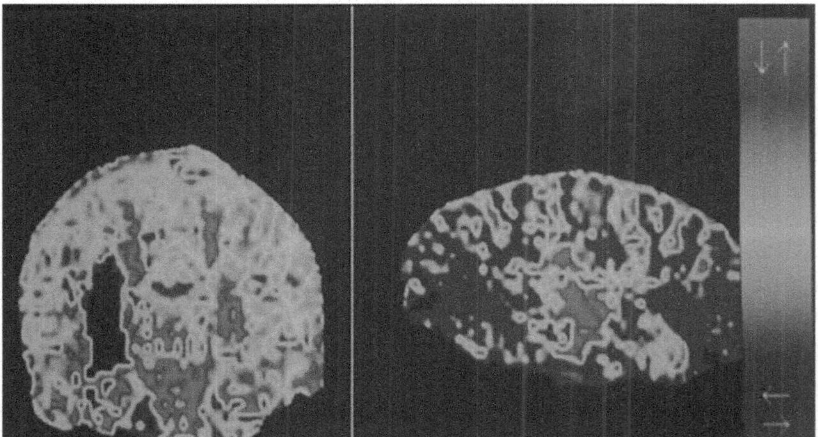

Fig. 1.10 The first color maps of white matter fiber orientation using diffusion MRI, 6 January, 2012 (*Source* From Wikimedia Commons, licensed under CC0)

has enabled the creation of bespoke experiences that cater to the individual sensory profiles of users. Such innovations in the digital realm, including applications and immersive environments, are revolutionizing how people with different neurological conditions interact with and perceive the world. These personalized digital experiences, which may include elements like slow lead-ins or the soothing sounds of a beach, are instrumental in meeting the specific sensory needs of individuals, including those with cognitive disabilities (Xiao & Stibor, 2010). The ongoing research into perceptual synthesis has unveiled the remarkable diversity in sensory perception and processing among individuals, influenced by factors such as language, cultural context, and genetics (Brang & Ramachandran, 2011; Brogaard & Sørensen, 2023). This understanding is crucial in helping individuals discover their *neurotribe*.

Journalist Steve Silberman introduced the term in his 2015 book *NeuroTribes: The Legacy of Autism and the Future of Neurodiversity*. The term signifies a pivotal change in our understanding of neurological diversity, with a particular focus on autism. It suggests that individuals who are neurodivergent, including those on the autism spectrum, are part of a distinct community or "tribe," each with its own set of unique traits, strengths, and ways of interacting with the world. This idea promotes a shift toward greater acceptance and appreciation of neurodiversity, moving away from perceiving neurodivergent conditions as mere disorders or deficits. In the context of synesthesia, the concept of a neurotribe is especially relevant, underscoring the distinct sensory experiences and perceptions that characterize these various groups of synesthetes. Used broadly in this book to refer to various neurotypes, including neurotypicals and neurodivergents, as well as synesthetes, neurotribes—communities with diverse methods of sensory information interpretation—can be leveraged to help individuals learn more about their own sensory processing, and identify procedures for sensory integration management. In essence, by understanding how an individual interprets sensory stimuli and how these stimuli evoke positive or negative sensations, emotions, and feelings, we can identify aesthetic interventions to create tailored experiences aimed at achieving specific outcomes. Consequently, this book aims to broaden the comprehension of neurodiverse sensory experiences across different neurotribes, enhancing the recognition and appreciation of the intricate and varied aspects of human cognition and perception.

Exploring neurotribes involves understanding how individuals with various neurological profiles interact with and respond to their environments, particularly in relation to sensory stimuli. Professionals in the fields of art and mental health are increasingly recognizing the need for a more targeted approach in addressing the distinct needs of different neurotypes (Quadt et al., 2023). This specificity in approach can elicit varied responses based on carefully tailored exposure to art, including exhibition design and collaboration with neurodivergent artists. In this context, it is more apt to refer to adjustments made for neurodivergent individuals as "success enablers" rather than mere accommodations (Dreaver et al., 2020). These enablers can be strategies within neuroarts designed to help neurodivergent individuals discover their neurotribe, better understand themselves, and enhance their quality of life. Inclusive art experiences play a pivotal role here, considering how art impacts the psychophysical aspects of humans (Antonelli et al., 2020).

Therefore, understanding commonalities between similar neural profiles, such as sensory processing disorders, is crucial and involves examining personal senses and changes—what specific stimuli like touch, smell, sound, or images activate an individual, and whether a connection between sensory preferences and visual stimuli exists. These insights can help individuals identify triggers and learn how to manage them, leading to self-realization and improved self-awareness. At the same time, the digital sphere, with its capacity for customization and control, offers unique opportunities for personalization to cater to the diverse needs of the neurodivergent mind. This includes tools and applications that allow users to experience the world through various sensory conditions, such as synesthesia, sensory processing times, or color blindness.

This study focuses on deepening our understanding of neurodiversity, aiming to help people identify their unique neurological group and better understand how they process sensory information and think. Moving beyond one-size-fits-all solutions, the research celebrates the complex variety found in human neurodiversity. Using a comprehensive approach that includes various neurodiverse groups, such as people with synesthesia, the study employs both surveys and one-on-one interviews to delve into the diverse experiences of these individuals. The data gathered is analyzed to develop new ways of determining one's neurotribe (Hutson & Hutson, 2023). Furthermore, the research investigates innovative digital methods, like using AI and immersive XR technology, to create personalized experiences that cater specifically to each individual's unique sensory and cognitive style. The results from this study are essential, as they guide the discussions in the book and provide a base for examining specialized interventions and strategies that align with the varied sensory needs and preferences of the participants of the study.

In the study examining neurodiversity, the demographic overview reveals a diverse group of participants who identify with various neurotypes (Table 1.1). Of the 94 individuals who took part, a small percentage, 6.38%, reported having dyslexia, amounting to six people. Only one participant, representing 1.06% of the sample, identified with dysgraphia. The most common neurotype among the respondents was ADHD, with 22 individuals, or 23.40% of participants, reporting this condition. PTSD was present in 11.70% of the respondents, totaling 11 individuals. Autism Spectrum Disorder (ASD) was identified by 9.57% of the group, which is nine participants. In addition to these specific conditions, 12.77% of the participants, or 12 individuals, specified other neurotypes that were not listed. There were ten individuals, making up 10.64% of the sample, who identified simply with autism, while Asperger's syndrome was reported by 5.32%, which is five participants. Notably, 19.15% of respondents, amounting to 18 individuals, identified as neurotypical, indicating no neurodivergent condition. This array of neurotypes among the participants underscores the spectrum of neurodiversity present within the population.

Table 1.1 Self-identified neurotypes

Neurotype	*Percentage (%)*	*Count*
Childhood disintegrative disorder	0.00	0
Kanner's syndrome	0.00	0
Dyslexia	6.38	6
Dysgraphia	1.06	1
ADHD	23.40	22
PMDD	0.00	0
PTSD	11.70	11
Autism Spectrum Disorder (ASD)	9.57	9
Other (please specify)	12.77	12
Autism	10.64	10
Pervasive developmental disorder (PDD-NOS)	0.00	0
Asperger's syndrome	5.32	5
No/neurotypical	19.15	18

The study also collected demographic information to understand the variety of participants involved. In terms of age, the participants spanned a range from 18 to over 65, with the largest group being those aged 35–44, representing 26.56% of the sample. The age groups of 25–34 and 45–54 were also well represented, with 20.31% and 15.63% respectively. The study included a younger demographic of 18–24-year-olds and an older demographic of 55–64-year-olds, each making up 14.06% of the participants, while those 65 or older comprised the smallest group at 4.69%. Additionally, 4.69% of participants chose not to disclose their age. Regarding gender identity, a significant majority of respondents identified as female, making up 75% of the participants. Males accounted for 23.44%, and a smaller fraction, 1.56%, identified as genderqueer/genderfluid. There were no participants who identified as non-binary, transgender, or other specified genders. When asked about Hispanic/Latinx identity, the vast majority of respondents, 93.44%, indicated they were not Hispanic/Latinx, while a small minority of 6.56% affirmed this cultural identity.

As for race and ethnic heritage, a substantial majority, 85.07%, identified as White/Caucasian. Other represented demographics included American Indian or Alaskan Native, Asian, and Black or African-American, each comprising 2.99% of participants. There were no participants who identified as Native Hawaiian or Pacific Islander. An additional 5.97% identified as belonging to other racial or ethnic groups not listed in the survey options. In the "Other" category specified by participants, responses included European, Mixed White Asian, mixed, and Middle Eastern heritage.

Additionally, the demographic data from the study provides insights into the sexual orientation, educational background, primary language, and physical disability status of the participants. Regarding sexual orientation, the majority of participants, 72.31%, identified as heterosexual/straight. Bisexual individuals accounted for 10.77%, while a smaller percentage identified as homosexual/gay/lesbian (1.54%), asexual (1.54%), and pansexual (3.08%). Queer individuals represented 6.15% of the sample, and 4.62% preferred not to disclose their sexual orientation.

In terms of education, the participants were well-educated, with the largest percentage holding a bachelor's degree (32.81%) and nearly an equal number possessing a master's degree (29.69%). Doctoral degree holders made up 10.94%, while some college education without a degree was reported by 12.50%. A few had an associate degree (7.81%), and minimal numbers had only a high school diploma or GED (3.13%) or did not have a high school diploma or GED (1.56%). Only 1.56% chose not to specify their educational level. The primary language spoken at home for the majority of participants was English, at 87.50%. The remaining responses included other languages such as Italian, French, and German, with some households speaking multiple languages. A small number spoke Portuguese (3.13%) or Arabic (1.56%).

Finally, within the realm of demographics, when asked about physical disabilities, the majority of respondents, 75.00%, reported having no disability. However, there were individuals with visual impairments (7.81%), mobility or physical disabilities (9.38%), and hearing impairments (1.56%). A few participants, 6.25%, specified other disabilities, which included conditions such as interstitial cystitis, lower back issues, vertigo, being overweight, which they considered a disability, and erythromelalgia.

The respondents of the study were also asked about their frequency of engaging with art in various settings. When inquired about their visits to museums or art galleries, a small minority, 1.56%, stated they never visit, while 26.56% said they rarely do, which equates to once a year or less. A significant portion of the participants, 34.38%, visit occasionally, defined as 2–4 times a year. Interestingly, the largest group, 37.50%, reported they frequently visit museums or art galleries, doing so five or more times a year.

Regarding the consumption of art digitally, the pattern of engagement is even more pronounced. A mere 1.56% of participants never view art on desktops, tablets, or mobile devices for enjoyment. Weekly art viewers make up 10.94%, and those who engage with art 2–3 times a week represent 17.19% of the sample. A slightly smaller group, 14.06%, views art 4–6 times a week. The majority of respondents, 56.25%, reported that they view art on their devices daily. These findings suggest that the participants are generally quite engaged with art, with a notable inclination toward integrating art into their daily lives through digital means. This high level of digital engagement points to the potential for online platforms and technologies to play a significant role in delivering personalized art experiences tailored to individual preferences and neurotypes.

The participants of the study provided further insight into their emotional and sensory engagement with art. When asked about the frequency of strong sensory reactions to art, such as goosebumps, tingling, or emotional overwhelm, a small minority of 3.13% stated they never have such experiences. The majority, 64.06%, sometimes experience these reactions, while about 10.94% feel them about half of the time they engage with art. A notable 14.06% have these strong sensory reactions most of the time, and 7.81% always have them when engaging with art. Additionally, when neurodivergent participants were asked whether their neurodivergence influences their perception or engagement with art, half of them (50%) responded affirmatively, stating that their neurodivergence impacts their experience with art in a positive way. Only a small fraction (1.67%) felt their neurodivergence had a negative influence, while 5% reported no discernible influence. Some participants, 16.67%, were unsure of the influence, and 26.67% stated the question was not applicable to them.

The study further explored the phenomenon of sensory "crossovers" or synesthesia among participants. When asked if they ever experience synesthesia, such as hearing colors or seeing sounds, 9.38% of respondents reported always experiencing these sensory crossovers. Another 15.63% often have such experiences, while 26.56% sometimes do. Rare occurrences were reported by 17.19%, and 31.25% never experience synesthetic sensations. Participants detailed various synesthetic experiences (Table 1.2): 12.86% see specific colors when hearing sounds, 4.29% taste flavors in response to words or numbers, and 11.43% feel tactile sensations when observing visual patterns. Other reported synesthetic experiences included perceiving days of the week or months as colors (8.57%) and associating letters or numbers with personalities or emotions (4.29%).

Within the "Other" category, participants in the study reported a variety of unique synesthetic experiences, illustrating the diverse ways in which synesthesia can manifest. One individual described "Mirror-touch" synesthesia, an intriguing condition where observing touch can evoke tactile sensations on one's own body. Another participant noted an affinity for certain textures and a vivid visual memory, suggesting a potential link to synesthetic perception that they had only recently begun to consider.

Some respondents reported multifaceted synesthetic experiences; for instance, one person mentioned hearing pain, a rare and intimate crossover of senses. Another shared that while the survey allowed for only one option to be selected, they experienced a range of synesthetic reactions, particularly during activities such as painting or cloud watching, with the color blue eliciting the strongest response.

Table 1.2 Reported synesthetic experiences

Sensation/perception	*Percentage (%)*	*Count*
Seeing specific colors when hearing certain sounds	12.86	9
Not applicable/none	28.57	20
Tasting flavors in response to specific words or numbers	4.29	3
Feeling tactile sensations when observing visual patterns	11.43	8
Associating specific shapes with particular tastes	1.43	1
Perceiving days of the week/months as distinct colors	8.57	6
Experiencing temperatures when hearing specific musical notes	1.43	1
Associating letters/numbers with personalities or emotions	4.29	3
Visualizing time units as spatial patterns/objects	4.29	3
Associating pitches/musical notes with distinct colors	2.86	2
Feeling textures when exposed to particular scents	2.86	2
Other (please specify)	17.14	12

Other unique associations described by participants included linking certain smells with specific images, albeit not direct associations like one might expect from a visual cue, but rather connections tied to memory. There were also mentions of sounds inducing the visualization of shapes, odors prompting visual imagery, and certain days, numbers, or letters being perceived in distinct colors. A few individuals described synesthetic experiences involving multiple senses, such as sounds and visuals evoking specific scents, or tasting flavors when exposed to words or numbers. Some reported that their sensory experiences went beyond simple perception, affecting emotions and physical sensations, like feeling a sound or seeing an emotion.

Moreover, there was a creative aspect to some responses, with participants curating soundscapes and visuals, suggesting that for them, synesthesia not only alters sensory perception but also informs their artistic expression. These personal accounts of synesthesia provide a deeper understanding of the condition, highlighting its complexity and the profound effect it can have on one's perception of the world. Additionally, participants were asked if they associate numbers or letters with specific colors, tastes, or smells. A few, 4.69%, always make such associations, 15.63% often do, and 10.94% sometimes do. However, the majority, 51.56%, never make these associations.

The responses also indicated varying degrees to which words or phrases evoke textures or temperatures, with 4.69% always experiencing this, 17.19% often, and 21.88% sometimes. A smaller group, 14.06%, rarely experience this, and 42.19% never do. Regarding the perception of time as a spatial dimension, 18.75% of participants always visualize time in this way, 20.31% often do, and 17.19% sometimes have this experience. Fewer, 15.63%, rarely experience time as a spatial dimension, and 28.13% never do. When considering the impact of synesthesia on daily life, 25.93% of respondents believe it enhances memory and recall. Some find that it makes everyday experiences more vivid (7.41%) or facilitates creative thinking (9.26%). Others face challenges, such as confusion or overstimulation (9.26%) or avoidance of certain stimuli (1.85%). However, 16.67% report an enhanced appreciation of music, art, or other sensory experiences.

Participants also shared their preferences in art, with abstract art attracting 21.54% of them. Others preferred realistic (16.92%), mixed media (16.92%), or digital art forms (18.46%). Some specified other preferences, including Impressionism and modern art. Lastly, respondents were asked about their emotional responses to art. A notable 26.56% feel their reactions are significantly more intense than others, while 37.50% believe theirs are somewhat more intense. A fair number, 18.75%, feel their emotional reactions are comparable to those of others, with only 1.56% feeling less intense reactions.

In the final open-response question of the survey, participants shared personal reflections on their experiences with art, revealing the intricate relationship between their neurodivergence and artistic engagement. One participant, a photographer, identified as an artist in their own right, suggesting a direct and intimate involvement with the creative process. Another respondent attributed their love for music, particularly electronic music, to ADHD/neurodivergence, describing how the layered complexity of this genre "tickles the brain" and provides the necessary stimulation to maintain focus, acting as a counterbalance to the search for distractions. Art-making was described as expansive and spatial, an experience that one participant felt contrasted with conventional education. There was a call for more representation and celebration of Neurodivergent artists within the art community. A struggle to discern personal perception from the artist's expression was noted, indicating the deeply subjective nature of art interpretation. One respondent expressed a profound connection to the field, revealing an ambition to pursue a

Doctorate in Art Therapy to explore art's physiological impact on the body. This suggests a desire to understand and bridge the gap between sensory experiences and therapeutic outcomes.

Color was mentioned as having spiritual associations, with the color blue being a recurrent theme during prayer and profound insights. Visual imagery, while not hallucinatory, was described as a powerful tool for processing experiences and navigating challenging times. The visual overstimulation provided by a plethora of colors and textures was celebrated by one participant, indicating a preference for rich, sensory-engaging environments. Conversely, another respondent described the challenges of emotional reading and responding, noting that art often instills a sense of peace and a desire for contemplation, regardless of the artwork's inherent emotion. The sound associated with visual art was another crossover experience shared, adding another dimension to the sensory engagement with art. A participant expressed a wish to provide multiple responses to survey inquiries, reflecting the complex and multifaceted nature of their experiences.

Finally, one individual pondered the creative process, questioning the genesis of artistic expression—whether it is the emotion that ignites creation or the act of creating that fosters emotion. They contemplated the language of art, its emergence or pre-construction, and the translatability of artistic expression. In all, these responses, rich with personal insight, underscore the diversity of experiences and the profound impact that neurodivergence can have on the perception and appreciation of art. They offer a valuable perspective on how art can serve as a language of expression and an avenue for connection for those with varied sensory and cognitive experiences.

The remainder of the treatment will contextualize these aggregate results and look more closely at the alignment of different neurotypes with sensory experiences. As such, the exploration continues in the next chapter with a deep dive into the integration of multimodal sensory modalities in art and technology. This exploration includes case studies of artists who have ingeniously incorporated sensory experiences into their work and discusses cutting-edge technology that enables readers to replicate and personalize these artistic experiences according to their sensory preferences. Following this, the book delves into the diverse classifications and types of sensory profiles, examining the current technological gaps in supporting these profiles. It discusses how these gaps can be bridged, paving the way for more personalized and inclusive sensory experiences.

Building on these insights, the book introduces a novel set of diagnostic criteria for identifying neurotribes. This section includes real-world case studies from the research conducted for this book, featuring examples of synesthetes and individuals with various sensory processing conditions. These case studies illuminate the complexity and diversity of sensory experiences, contributing to a better understanding of neurodiversity.

The concluding part of the book looks toward the future of neuroarts. It proposes various strategies based on existing technology and research to address the needs of neurodiverse individuals. This includes the development of new bespoke experiences that allow individuals to identify their sensory profiles and preferences. Utilizing AI, the book envisions creating immersive experiences in XR (Extended Reality) tailored to evoke desired moods, emotions, or feelings, thus enriching the sensory world of each individual. Through this exploration, the book aims not only to expand our understanding of neurodiversity but also to foster a deeper appreciation for the complex and diverse nature of human cognition and perception.

1.7 Summary of Chapters

1.7.1 Chapter 2: Integrating Sensory Modalities and Technologies in Artistic Contexts

This chapter delves into the innovative fusion of sensory technologies and art, illustrating how this synergy is revolutionizing artistic expression and experience. It begins by examining sensory technologies as new mediums in the arts, focusing on ambient sound applications and the transformative impact of virtual and augmented reality in visual arts. This includes exploring virtual tourism and its influence on cultural aesthetics, the use of VR in language revitalization, and how AR is reshaping color interpretation in art. The chapter then presents a series of artist case studies, showcasing how visual, auditory, tactile, olfactory, and gustatory technologies are being integrated into artistic processes. These studies highlight solutions for colorblindness, immersive language learning environments, auditory representations of visual phenomena, and the use of digital scent technology. The future of sensory experience in art is also explored, discussing the roles of projected synesthesia in art therapy, multisensory technologies in bridging sensory gaps, and emerging trends like tangible media and biofeedback in immersive art experiences. This chapter not only illustrates current innovations but also forecasts the exciting potential of sensory technologies in expanding the boundaries of how we create, interact with, and perceive art in the modern world.

1.7.2 *Chapter 3: Sensory Profiles and Technological Deficits: Classification, Relevance, and Gaps*

The chapter provides an analysis of the diverse range of sensory experiences and the challenges in addressing them with current technology. Opening with a discussion on how cultural and spiritual beliefs shape sensory perceptions, the chapter uses insights from various cultures, including the Berinmo people. It then delves into various aesthetic disabilities beyond visual perception, such as face blindness, dyschromatopsia, dyscalculia, and synesthesia, exploring their unique impacts on artistic interpretation. The chapter also tackles cross-modal sensory challenges, highlighting conditions like anosmia and agnosia, and their implications for experiencing art. Emphasizing neurodiversity, it examines how neurodivergent individuals, particularly those with autism, have distinct sensory preferences and responses to art. Aphantasia's role in artistic experience is discussed, exploring how the absence of mental imagery influences art perception and creation. Finally, the chapter identifies key gaps in current technological applications, discussing issues like ableism in sensory technologies and the need for more inclusive and adaptable tools. It concludes by exploring therapeutic and adaptive applications, including color therapy and sensory adaptations in museums, emphasizing the importance of developing technologies that cater to a broad spectrum of sensory experiences and preferences. This chapter provides crucial insights into the intricate relationship between sensory experiences, neurodiversity, and the role of technology in enhancing artistic engagement.

1.7.3 *Chapter 4: Diagnostics in Neurotribe Identification: Case Studies*

This chapter navigates the complex world of neurodiversity in relation to artistic expression and perception. It begins by establishing a theoretical foundation, exploring enactivism and phenomenology to understand how sensory experiences are shaped by cognitive processes and personal realities. This foundation is brought to life through case studies and interviews that demonstrate these concepts in practical scenarios. The chapter then delves into Sensory Processing Sensitivity (SPS), exploring how it affects aesthetic responses and is expressed in artistic interactions. It examines the intricate link between aesthetic preferences, emotional responses, and their underlying neurological mechanisms, offering insights gained from interviews about emotional reactions to art. Diagnostic approaches, such as the use of art diaries and pattern identification in artistic preferences, are discussed to understand sensory profiles better. The chapter also addresses the origins and impacts of sensory biases from infancy to adulthood, highlighting the diversity in artistic preferences among neurodivergent individuals, especially focusing on autism and sensory processing disorders. It explores the role of technology in aiding diagnosis and profiling, including AI-driven approaches and app-based neurotype tracking. The complex interplay of sensory experiences, such as synesthesia, and their influence on art perception are also covered. Concluding with future directions, the chapter points to emerging research areas like personalized quizzes and reduced chromoscapes, signaling new avenues in the study of neurodiversity and art.

1.7.4 *Chapter 5: Conclusion: Future Applications, Technologies, and Directions*

The conclusion serves as a forward-looking summary of the evolving field of neuroarts. It discusses the potential for personalization in neuroaesthetic experiences through sensory augmentation and advanced neurological technologies. The chapter explores future applications, including virtual art in therapy, AI's role in sensory enhancement, and the use of tactile therapies for multisensory artistic engagement. It also delves into neurophysiological stimulation in art, such as affective computing and cognitive techniques, and examines the intersection of AI with artistic exploration, particularly in creating immersive, virtual experiences. Ethical considerations in neuroarts, including data privacy and the impact of commercialization, are highlighted, emphasizing the need for responsible development. The chapter underscores the importance of public engagement and the integration of neuroarts in education and healthcare, suggesting its potential as a therapeutic tool. It concludes by looking at the technological frontiers in neuroarts, like brain–computer interfaces and machine learning in art creation, and addresses theoretical questions about subjectivity and realism in art. Ultimately, this chapter paints a picture of neuroarts as a field rich with possibilities, set to redefine our understanding and engagement with art through the lens of neuroscience and technology.

References

Allen, J. P. (2005). *The art of medicine in ancient Egypt*. Metropolitan Museum of Art.

Antonelli, M., Donelli, D., Barbieri, G., Valussi, M., Maggini, V., & Firenzuoli, F. (2020). Forest volatile organic compounds and their effects on human health: A state-of-the-art review. *International Journal of Environmental Research and Public Health, 17*(18), 6506.

Bahrick, L. E., & Lickliter, R. (2014). Learning to attend selectively: The dual role of intersensory redundancy. *Current Directions in Psychological Science, 23*(6), 414–420.

Bennett, T. (2023). Indigenous art and the contemporary Australian art field. *Global Art in Local Art Worlds: Changing Hierarchies of Value*, 2.

Bigliassi, M. (2022). Functional significance of the dorsolateral prefrontal cortex during exhaustive exercise. *Biological Psychology, 175*, 108442.

Birnbaum, L. C. (1955). Behaviorism in the 1920's. *American Quarterly, 7*(1), 15–30.

Bouvet, L., Donnadieu, S., Valdois, S., Caron, C., Dawson, M., & Mottron, L. (2014). Veridical mapping in savant abilities, absolute pitch, and synesthesia: An autism case study. *Frontiers in Psychology, 5*, 106.

Bouvet, L., Amsellem, F., Maruani, A., Dupont, A. T. V., Mathieu, A., Bourgeron, T., Delorme, R., & Mottron, L. (2019). Synesthesia & autistic features in a large family: Evidence for spatial imagery as a common factor. *Behavioural Brain Research, 362*, 266–272.

Brang, D., & Ramachandran, V. S. (2011). Survival of the synesthesia gene: Why do people hear colors and taste words? *PLoS Biology, 9*(11), e1001205.

Brogaard, B., & Sørensen, T. A. (2023). Perceptual variation in object perception: A defence of perceptual pluralism. *Sensory Individuals: Unimodal and Multimodal Perspectives*, 113.

Burge, G. M. (2015). *A week in the life of a Roman Centurion*. InterVarsity Press.

Case-Smith, J., & Arbesman, M. (2008). Evidence-based review of interventions for autism used in or of relevance to occupational therapy. *American Journal of Occupational Therapy, 62*(4), 416–429.

Chapin, M., & Fish, B. (2016). The therapist as artist. In *Approaches to art therapy* (pp. 33–49). Routledge.

Chatterjee, A., & Vartanian, O. (2014). Neuroaesthetics. *Trends in Cognitive Sciences, 18*(7), 370–375.

Cheung, M. C., Law, D., Yip, J., & Wong, C. W. (2019). Emotional responses to visual art and commercial stimuli: Implications for creativity and aesthetics. *Frontiers in Psychology, 10*, 14.

Clary, P. L. (2010). Poetry and healing at the end of life. *Journal of Pain and Symptom Management, 40*(5), 796–800.

Colino, S. (2023, October 9). Beyond white noise: How different 'color' sounds help or hurt. *The Washington Post*. https://www.washingtonpost.com/home/2023/10/09/white-noise-color-sounds-brown-pink/

Cotter, K. N., Harrouche, M., Rodriguez-Boerwinkle, R. M., Boerwinkle, M., Silvia, P. J., & Pawelski, J. O. (2023). Virtual art visits: Examining the effects of slow looking on well-being in an online environment. *Psychology of Aesthetics, Creativity, and the Arts*. https://doi.org/10.1037/aca0000548.

Cytowic, R. E. (1995). Synesthesia: Phenomenology and neuropsychology. *Psyche, 2*(10), 2–10.

Cytowic, R. (2018). *Synesthesia*. MIT Press.

Del Campo, M. A., & Kehle, T. J. (2016). Autonomous sensory meridian response (ASMR) and frisson: Mindfully induced sensory phenomena that promote happiness. *International Journal of School & Educational Psychology, 4*(2), 99–105.

De'Souza, C. (2012). *Art and neuroscience: The historical emergence and conceptual context of neuro-art*. Doctoral Dissertation. School of Humanities and Social Science, Jacobs University, Bremen.

Dreaver, J., Thompson, C., Girdler, S., Adolfsson, M., Black, M. H., & Falkmer, M. (2020). Success factors enabling employment for adults on the autism spectrum from employers' perspective. *Journal of Autism and Developmental Disorders, 50*, 1657–1667.

Disney's, W. (2023). Every beautiful sound also creates an equally beautiful picture. *Lowering the Boom: Critical Studies in Film Sound*, 183.

Dwyer, P. (2022). The neurodiversity approach(es): What are they and what do they mean for researchers? *Human Development, 66*(2), 73–92.

Engel, G. L. (1977). The need for a new medical model: A challenge for biomedicine. *Science, 196*(4286), 129–136.

Fancourt, D., & Finn, S. (2019). *What is the evidence on the role of the arts in improving health and well-being? A scoping review*. World Health Organization. Regional Office for Europe.

Ferrari, G. R. F. (2019). Aristotle on musical catharsis and the pleasure of a good story. *Phronesis, 64*(2), 117–171.

Garzón Chiriví, O. A. (2015). Approach to a state of the art of traditional and popular medicinal practices in Hispano-American. *Folios, 41*, 157–168.

Golden, T. (2023, September 13). Arts on prescription: An emerging model of mental healthcare. *Psychology Today*. https://www.psychologytoday.com/us/blog/the-art-effect/202309/arts-on-prescription-an-emerging-model-of-mental-healthcare

Golden, T. L., Maier Lokuta, A., Mohanty, A., Tiedemann, A., Ng, T. W., Mendu, M., Morgan, N., Kuge, M. N., & Brinza, T. (2023). Social prescription in the US: A pilot evaluation of Mass Cultural Council's "CultureRx." *Frontiers in Public Health, 10*, 1016136.

Glasser, S. (2023). The relationship between music-related types of synesthesia and mental imagery in synesthete musicians. *Music & Science, 6*, 20592043231173810.

Goodley, D. (2016). Disability studies: An interdisciplinary introduction. *Disability Studies*, 1–296.

Greenberg, M. (2009, September). Synesthesia and literary symbolism. In *Forum Italicum* (Vol. 43, No. 2, pp. 362–384). Sage.

Grossenbacher, P. G., & Lovelace, C. T. (2001). Mechanisms of synesthesia: Cognitive and physiological constraints. *Trends in Cognitive Sciences, 5*(1), 36–41.

Hoffman, S. N., Urosevich, T. G., Kirchner, H. L., Boscarino, J. J., Dugan, R. J., Withey, C. A., Adams, R. E., Figley, C. R., & Boscarino, J. A. (2019). Grapheme-color synesthesia is associated with PTSD among deployed veterans: Confirmation of previous findings and need for additional research. *International Journal of Emergency Mental Health, 21*(1).

Hubbard, E. M. (2007). Neurophysiology of synesthesia. *Current Psychiatry Reports, 9*(3), 193–199.

Hulbig, P. R. (2023). To be neurodivergent. In *The pedagogy of self-authorship: The neurocognitive impact of science and metacognition* (pp. 181–191). Springer Nature Switzerland.

Hutson, J., & Hutson, P. (2023). Sensemaking and neuroaesthetics study. Lindenwood University.

Jewanski, J., Simner, J., Day, S. A., & Ward, J. (2011). The development of a scientific understanding of synesthesia from early case studies (1849–1873). *Journal of the History of the Neurosciences, 20*(4), 284–305.

Jewanski, J., Day, S. A., Simner, J., & Ward, J. (2013). The beginnings of an interdisciplinary study of synaesthesia: Discussions about the Nussbaumer brothers (1873). *Theoria et Historia Scientiarum, 10*, 149–176.

Junge, M. B. (2015). History of art therapy. *The Wiley handbook of art therapy*, 7–16. https://doi.org/10.1002/9781118306543.ch1

Kandinsky, W. (2012). *Concerning the spiritual in art*. Courier Corporation.

Lewkowicz, D. J. (2014). The development of intersensory temporal perception: An epigenetic systems/limitations view. *Psychological Bulletin, 140*(2), 281–308.

Lupton, E., & Lipps, A. (Eds.). (2018). *The senses: Design beyond vision*. Chronicle Books.

Magsamen, S., & Ross, I. (2023). *Your brain on art: How the arts transform us*. Random House.

Magsamen, S., Golden, T. L., Towriss, C. A., & Allen, J. (2023). The impact thinking framework: A process for advancing research-to-practice initiatives in neuroaesthetics. *Frontiers in Psychology, 14*, 1129334.

Marco, E. J., Hinkley, L. B., Hill, S. S., & Nagarajan, S. S. (2011). Sensory processing in autism: A review of neurophysiologic findings. *Pediatric Research, 69*(5 Pt 2), 48R-54R.

Mastandrea, S., Fagioli, S., & Biasi, V. (2019). Art and psychological well-being: Linking the brain to the aesthetic emotion. *Frontiers in Psychology, 10*, 739.

Macnaughton, J., White, M., & Stacy, R. (2005). Researching the benefits of arts in health. *Health Education, 105*(5), 332–339.

Mather, G. (2013). *The psychology of visual art: Eye, brain and art*. Cambridge University Press.

Michels, K., Dubaz, O., Hornthal, E., & Bega, D. (2018). "Dance Therapy" as a psychotherapeutic movement intervention in Parkinson's disease. *Complementary Therapies in Medicine, 40*, 248–252.

Miller, L. J., Anzalone, M. E., Lane, S. J., Cermak, S. A., & Osten, E. T. (2007). Concept evolution in sensory integration: A proposed nosology for diagnosis. *American Journal of Occupational Therapy, 61*(2), 135–140.

Nanay, B. (2020). Synesthesia as (multimodal) mental imagery. *Multisensory Research, 34*(3), 281–296.

Nerurkar, N. K., Chitnis, T. A., & Pereira, J. (2022). Presence and pattern of chromesthesia in 200 individuals: An experiment performed on World Voice Day. *International Journal of Phonosurgery & Laryngology, 12*(1), 12–15.

Neufeld, J., Roy, M., Zapf, A., Sinke, C., Emrich, H. M., Prox-Vagedes, V., Dillo, W., & Zedler, M. (2013). Is synesthesia more common in patients with Asperger syndrome? *Frontiers in Human Neuroscience, 7*, 847.

Nowak, R., & Gimenez, S. (2015). Magnetoencephalography (MEG) and diffusion tensor magnetic resonance imaging (DTI) of functional and effective brain connectivity in traumatic brain injury. *Clinical Neurophysiology, 127*(3), 75.

Pallasmaa, J. (2012). *The eyes of the skin: Architecture and the senses*. Wiley.

Puranik, V. (2023, November 14). Looking at colorful art helps our brain improve its well-being. *Medicinal Media*. https://www.medicinal-media.org/explore/colorful-art-helps-the-brain

Oliver, M. (1996). *Understanding disability: From theory to practice*. Macmillan International Higher Education.

Quadt, L., Williams, G., Mulcahy, J., Larsson, D. E., Silva, M., Arnold, A. J., Critchley, H., & Garfinkel, S. N. (2023). "I'm trying to reach out, I'm trying to find my people": A mixed-methods investigation of the link between sensory differences, loneliness, and mental health in autistic and nonautistic adults. *Autism in Adulthood*. https://doi.org/10.1089/aut.2022.0062

Raad, G., Tanios, J., Azoury, J., Daher, A., Fakih, C., & Bakos, H. W. (2021). Neurophysiology of cognitive behavioural therapy, deep breathing and progressive muscle relaxation used in conjunction with ART treatments: A narrative review. *Human Reproduction Update, 27*(2), 324–338.

Radulski, E. M. (2022). Conceptualising autistic masking, camouflaging, and neurotypical privilege: Towards a minority group model of neurodiversity. *Human Development, 66*(2), 113–127.

Riedel, A., Maier, S., Wenzler, K., Feige, B., Tebartz van Elst, L., Bölte, S., & Neufeld, J. (2020). A case of co-occurring synesthesia, autism, prodigious talent and strong structural brain connectivity. *BMC Psychiatry, 20*(1), 1–11.

Roe, A. W., Chelazzi, L., Connor, C. E., Conway, B. R., Fujita, I., Gallant, J. L., Lu, H., & Vanduffel, W. (2012). Toward a unified theory of visual area V4. *Neuron, 74*(1), 12–29.

Rouw, R., & Scholte, H. S. (2007). Increased structural connectivity in grapheme-color synesthesia. *Nature Neuroscience, 10*(6), 792–797.

Samuels, J. (1995). Healing sand: A comparative investigation of Navajo sandpaintings and Tibetan sand maṇḍalas. *Journal of Ritual Studies*, 101–126.

Schlich, E. B. (2012). *Not nobody, not nohow* (Doctoral dissertation, Bowling Green State University).

Secker, J., Heydinrych, K., Kent, L., & Keay, J. (2018). Why art? Exploring the contribution to mental well-being of the creative aspects and processes of visual art-making in an arts and mental health course. *Arts & Health, 10*(1), 72–84.

Shakespeare, T. (2014). *Disability rights and wrongs revisited*. Routledge.

Siler, T. (2015). Neuroart: Picturing the neuroscience of intentional actions in art and science. *Frontiers in Human Neuroscience, 9*, 410.

Silberman, S. (2015). *Neurotribes: The legacy of autism and the future of neurodiversity*. Penguin.

Singer, J. (1999). Why can't you be normal for once in your life? From a problem with no name to the emergence of a new category of difference. In M. Corker & S. French (Eds.), *Disability discourse* (pp. 59–67). Open University Press.

Singh, S., Sabahat, M., & Qamrudiin, J. (2021). The impact of architecture in the process of healing & well-being. *International Journal for Research in Applied Science & Engineering Technology, 9*(3), 202–222.

Smyth, N., Rossi, E., & Wood, C. (2020). Effectiveness of stress-relieving strategies in regulating patterns of cortisol secretion and promoting brain health. *International Review of Neurobiology, 150*, 219–246.

Surapaneni, K. M. (2023). Enriching anatomy education with the integration of rangoli: Nurturing cultural practices in medical education. *Medical Science Educator, 33*(5), 1293–1293.

Taylor, J. B. (2021). *Whole brain living: The anatomy of choice and the four characters that drive our life*. Hay House, Inc.

Thibault, G. E. (2019). Humanism in medicine: What does it mean and why is it more important than ever? *Academic Medicine, 94*(8), 1074–1077.

Uszerowicz, M. (2019, April 9). At Google design's a space for being, good aesthetics are good for the brain. *Cultured*. https://www.culturedmag.com/article/2019/04/09/google-design

Van Campen, C. (2010). *The hidden sense: Synesthesia in art and science*. MIT Press.

Van Leeuwen, T. M., Wilsson, L., Norrman, H. N., Dingemanse, M., Bölte, S., & Neufeld, J. (2021). Perceptual processing links autism and synesthesia: A co-twin control study. *Cortex, 145*, 236–249.

Von Bernhardi, R., Bernhardi, L. E. V., & Eugenín, J. (2017). What is neural plasticity? *The Plastic Brain* (pp. 1–15). https://doi.org/10.1007/978-3-319-62817-2_1.

Ward, J. (2013). Synesthesia. *Annual Review of Psychology, 64*, 49–75.

Ward, J. (2019). The co-occurrence of mirror-touch with other types of synaesthesia. *Perception, 48*(11), 1146–1152.

Whitaker, K. J., Kang, X., Herron, T. J., Woods, D. L., Robertson, L. C., & Alvarez, B. D. (2014). White matter microstructure throughout the brain correlates with visual imagery in grapheme–color synesthesia. *NeuroImage, 90*, 52–59.

Windle, G., Joling, K. J., Howson-Griffiths, T., Woods, B., Jones, C. H., Van de Ven, P. M., Newman, A., & Parkinson, C. (2018). The impact of a visual arts program on quality of life, communication, and well-being of people living with dementia: A mixed-methods longitudinal investigation. *International Psychogeriatrics, 30*(3), 409–423.

Xiao, H., & Stibor, T. (2010, December). Toward artificial synesthesia: Linking images and sounds via words. In *NIPS Workshop on Machine Learning for next generation Computer Vision challenges*.

Zeki, S. (1999). Splendours and miseries of the brain. Philosophical Transactions of the Royal Society of London. Series B: *Biological Sciences, 354*(1392), 2053–2065.

Zeki, S. (2007). The neurobiology of love. *FEBS Letters, 581*(14), 2575–2579.

Zhang, Y., Cai, J., An, L., Hui, F., Ren, T., Ma, H., & Zhao, Q. (2017). Does music therapy enhance behavioral and cognitive function in elderly dementia patients? A systematic review and meta-analysis. *Ageing Research Reviews, 35*, 1–11.

Zhang, Y. H., & Rose, K. (2001). *A brief history of qi*. Paradigm Publications.

PART II

Understanding Sensory Experiences

CHAPTER 2

Integrating Sensory Modalities and Technologies in Artistic Contexts

This chapter delves into the innovative fusion of sensory technologies and art, illustrating how this synergy is expanding artistic expression and experience. It begins by examining sensory technologies as new mediums in the arts, focusing on ambient sound applications and the transformative impact of virtual and augmented reality in visual arts. This includes exploring virtual tourism and its influence on cultural aesthetics, the use of VR in language revitalization, and how AR is reshaping color interpretation in art. The chapter then presents a series of artist case studies, showcasing how visual, auditory, tactile, olfactory, and gustatory technologies are being integrated into artistic processes. These studies highlight solutions for colorblindness, immersive language learning environments, auditory representations of visual phenomena, and the use of digital scent technology. The future of sensory experience in art is also explored, discussing the roles of projected synesthesia in art therapy, multisensory technologies in bridging sensory gaps, and emerging trends like tangible media and biofeedback in immersive art experiences. This chapter not only illustrates current innovations but also forecasts the exciting potential of sensory technologies in expanding the boundaries of how we create, interact with, and perceive art in the modern world.

J. Hutson et al., *Sensemaking and Neuroaesthetics*,
https://doi.org/10.1007/978-3-031-58045-1_2

2.1 Sensory Technologies as Artistic Medium

Exploring the intersection of art and sensory technology reveals a transformative space where traditional media merges with cutting-edge innovations. Historically, art has been experienced within the physical confines of galleries, museums, and performance spaces. Now, these boundaries have been transcended as art enters the spheres of immersive experiences using a variety of emergent technology, including extended reality. Extended reality (XR) represents a suite of technologies that enhance our perception of the digital and physical worlds. Augmented reality (AR) layers virtual elements over the real world, while virtual reality (VR) immerses users in a completely virtual environment. Mixed reality (MR) merges the two, allowing real and virtual elements to coexist and interact in a single space. Such novel progressions of immersion into multisensory dimensions offer audiences novel and profound ways to interact with artistic works (Vital et al., 2023). Thus, the exploration starts with visual enhancements, such as XR, which have become the forerunners in this evolution. For instance, VR allows for a form of tourism that transcends physical and geographical barriers, inviting viewers into digitally reconstructed spaces that pulse with cultural and historical resonance (Hutson & Olsen, 2022). AR layers digital information onto the physical world, altering and enriching perceptions of art, allowing even those with visual impairments to experience the kaleidoscope of colors and forms (Akın & Cömert, 2023).

Beyond the traditional forms of experiencing art, and in keeping with the challenge to ocularcentrism, artists transcend traditional visual-centric methods by engaging multiple senses in installations and interactive exhibits, fostering holistic sensory experiences. In considering other sense experiences, ambient soundscapes represent another realm for investigation with artists using them to envelop viewers in an auditory experience that complements the visual (Fraisse et al., 2023). This auditory dimension can stand alone as a form of art or blend with visual, tactile, or olfactory elements to create a holistic experience (Murao, 2022). Spatial-audio technology envelops the listener, creating a three-dimensional soundscape that can replicate the acoustics of a Medieval cathedral or the intimate whispers of a quiet, modern gallery.

While sight and sound in exhibition spaces may be familiar to the artgoing public, there are examples of other artists who challenge convention by tapping into cross-sensory perceptions. These less charted territories include tactile (touch), olfactory (smell), and even gustatory (taste) experiences within art. Tactile feedback systems bring texture and touch to digital creations, allowing objects (or even light and sound) to be felt as well as seen (Chaudhury et al., 2022). Olfactory technologies introduce scent as an element of narrative and ambiance, while gustatory experiments challenge artists and viewers to consider taste as a component of the aesthetic experience (Bembibre & Strlič, 2022). Looking ahead, a future where technology enables the creation of synesthetic experiences will allow individuals to perceive art in ways typically exclusive to those with the neurological condition of synesthesia. These experiences merge novelty with expanding the inclusivity of art and bridge gaps for those with different sensory abilities. Emerging technologies which allow for a dialogue between the physical responses of viewers and the art itself, like tangible media and biofeedback, create a potential feedback loop which personalizes and deepens the engagement (Castelló et al., 2023). Therefore, the boundaries of art are continuously being pushed further by technological advances, promising a future where artistic expression and sensory experience are limited only by the imagination. It heralds a time when the full spectrum of human sensory and emotional experience can be engaged and celebrated through art.

The fusion of art and sensory perception has given rise to multisensory exhibitions which mirror the layered experiences of synesthetes. Renowned and emerging artists alike are navigating these waters, bringing to life the intangible qualities of synesthesia in their work. The art of David Hockney (1937–), for instance, dissects the nuances of color perception (Fig. 2.1), while Christa Sommerer (1964–) creates interactive pieces that meld the organic with the digital (Bergantini, 2019; Boden & Edmonds, 2019). Similarly, French artist Joanie Lemercier (1982–) transports viewers into the heart of nature through her illuminated geometric patterns (https://joanielemercier.com/) (Tian, 2023). These forays into sensory fusion have been more thoroughly explored by Yann Marussich (1966–) (https://www.yannmarussich.ch/home.php) and Shiro Takatani (1963–) (http://shiro.dumbtype.com/works/index) in their performative works, exemplifying the intersection, merging sound, light, aroma, and texture to replicate the synesthetic experience in a setting meant to both disorient and captivate (Andrieu et al., 2020). Similarly, the installations of Anicka Yi (1971–) (https://www.anickayistudio.biz/) weave together scent and sound, creating an environment that engages all senses, while the works of Gustav Metzger (1926–2017) and Nam June Paik (1932–2006) present a dialogue on how technology shapes human experience (Barrett, 2023; Daris, 2021). A notable example is *TV Buddha* (1974) (Fig. 2.2) from Paik, a traditional Buddha statue is placed in a contemplative pose, facing a television that broadcasts a live feed of the statue itself. This innovative work melds technology with spirituality, presenting a thought-provoking intersection of Eastern philosophy and Western media culture. The arrangement creates a loop of reflection, where the Buddha, symbolizing peace and meditation, is juxtaposed with the dynamic, ever-changing nature of modern media. This continuous loop blurs the lines between reality and its digital representation, inviting viewers to ponder the interplay of the ancient and the contemporary, and the profound implications of this fusion on our perception of reality and spirituality (Lim, 2019).

The exhibit at the Museum of Modern Art (MoMA), *Sounds: A Contemporary Score* (2013) (https://www.moma.org/calendar/exhibitions/1351), brought the phenomenon of synesthesia into the public eye, featuring a group exhibition of sound art that allowed visitors to experience the intertwining of the senses firsthand (Heddaya, 2013). Emilie Gossiaux (1989–) (http://www.emiliegossiaux.com/), influenced by neuroscience, provides glimpses into the synesthetic process through her art (Lane, 2023). On the other hand, German-born artist Diemut

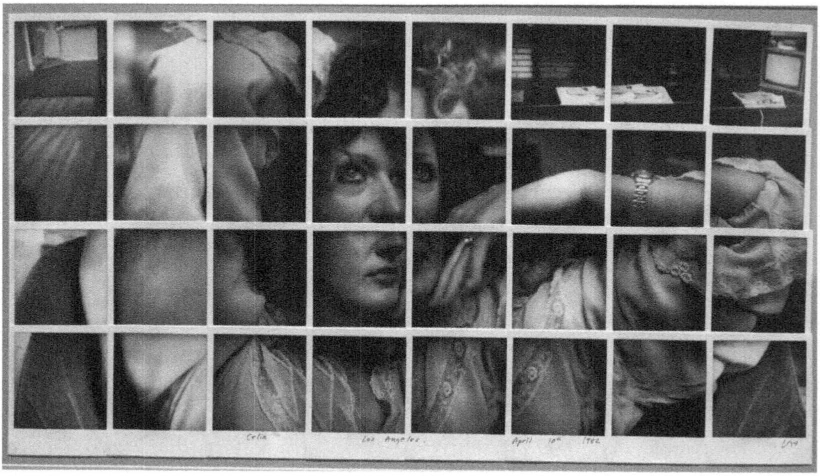

Fig. 2.1 David Hockney, *Celia*, Los Angeles. April 10, 1982 (*Source* from Wikimedia Commons, licensed under CC-BY 2.0)

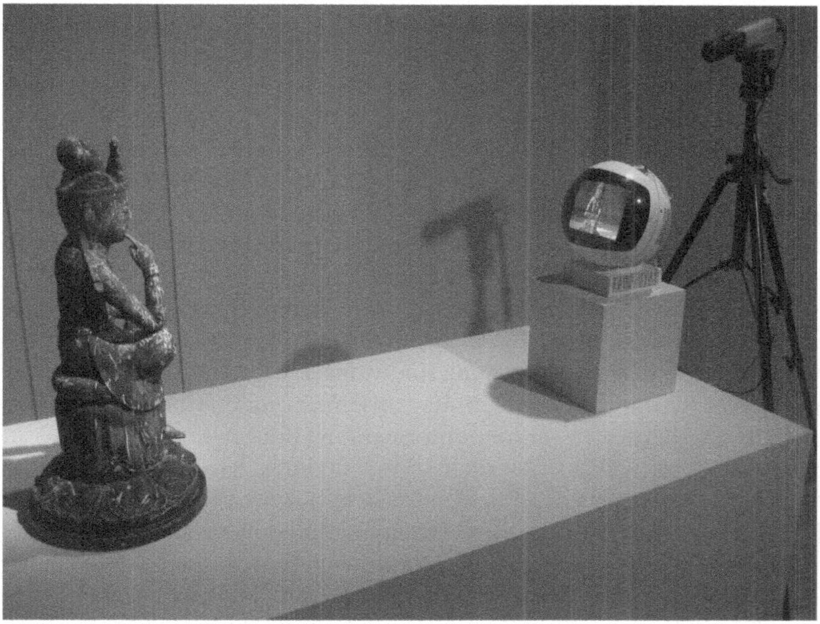

Fig. 2.2 Nam June Paik, *TV Buddha*, 1974 (*Source* from Wikimedia Commons, licensed under CC-BY 2.0)

Strebe (1982–) (https://www.diemutstrebe.com/) takes a more extreme approach in "Works on the Intersection of Art and Science," plunging visitors into sensory deprivation, challenging them to find meaning in the void (Sexton, 2021). These examples represent a thematic mosaic where art not only imitates life but also seeks to transcend ordinary sensory experiences, inviting a deeper exploration of perception and the human mind.

As these examples illustrate, artists are increasingly conscious of the diverse sensory profiles and conditions of audiences, creating works that acknowledge and embrace these differences. The artist-activist Liz Crow work, such as *Resistance* (2009), utilizes technology to simulate the perceptual experiences of individuals with disabilities, while the UK-based Tim Etchell (1962–) also delves into lived experience of physical disability in his performance art (Prendergast, 2021). Other examples include the sensory-rich installations of Stokholm-based Christine Ödlund

(1963–) (https://www.christineodlund.se/), which immerses viewers in a world akin to blindness, whereas conceptual artist-photographer Teresa Margolles (1963–) (https://macm.org/en/exhibitions/teresa-margolles-mundos/) employs forensic science to address themes of trauma and disability (Olivas, 2023). With each of these examples, various technologies are used to simulate different sensory experiences, primarily for a neurotypical and non-disabled audience. However, other creatives have sought to more extensively represent neurodivergent perspectives by and for these audiences, including autism.

This sensitivity to neurodiversity is bringing forward a surge in creativity aimed at neurological inclusivity. Evidence suggests a higher prevalence of synesthesia among individuals with autism, particularly mirror-touch synesthesia, which may arise from unique brain processing patterns in social and sensory information (Bouvet et al., 2019; Van Leeuwen et al., 2021). As such, these findings have inspired artistic endeavors that cater to and celebrate these distinct sensory experiences. The Reality Center (https://www.realitymgmt.com/about), for instance, has pioneered an example that merges enjoyment with therapeutic neuroscientific approaches, offering biofeedback and frequency technologies designed to stimulate the brain in ways akin to meditation or psychedelic therapies. In such a way, the center serves as a sanctuary for those who have experienced trauma, providing a rebomgulated environment free from overstimulation and promoting mental well-being without relying solely on traditional therapeutic conversations.

Aside from organizations, artists are also creating art that provides insights into the autistic experience. For instance, artists with autism, like Anna Berry, are creating works (https://www.annaberry.co.uk/) that center around experiences of marginalized groups, while Amy Sequenzia, a non-speaking autistic artist, employs painting and installation art to forge a visual language that expresses her thoughts and experiences beyond verbal communication (Brackett, 2022). Likewise, artists like Vietnamese-Australian visual artist and storyteller Matt Huynh (https://www.matthuynh.com/) and American visual artist Liz Phillips (https://lizphillips.net/w/) are contributing to this dialogue by focusing on how individuals with autism perceive sensory details. The drawings of Huynh emphasize sensory elements such as light and texture, while Phillips uses sound in her artwork to explore and communicate the sensory experiences of those with autism and sensory processing disorders (Cressman & DeTora, 2021). The experiences of neurodivergence are also vividly captured in multimedia works of Huynh such as *Reconfiguration: An Evening With Other Lives* (2015) (https://vimeo.com/145831948) Therefore, a growing number of artists are recognizing the sensory differences inherent in various neurotypes and are using this understanding in crafting their art to reflect, resonate with, and respect these differences. This evolving landscape of art underscores the potential for creative expression to serve as a powerful conduit for understanding, inclusivity, and awareness of the rich tapestry of human sensory and cognitive experiences.

The journey through the realm of multisensory art and its profound impact on perception and experience finds a complementary perspective in the Aesthetic Mindset Index (https://www.yourbrainonart.com/aesthetic-mindset-index) developed by computational cognitive neuroscientist Ed Vessel (Magsamen & Ross, 2023). This index, assessing aesthetic responsiveness and the influence of art on environmental attunement, adds depth to our understanding of how art affects our physical and psychological states and provides a valuable framework for comprehending the power of art to influence not only our physical responses but also our perception and interpretation of color within different cultural contexts. This exploration sets the stage for delving into specific case studies in visual technologies. These studies will further illuminate how technological advancements in visual arts are reshaping our engagement with art, enhancing our sensory experiences, and offering new insights into the intricate relationship between art, technology, and the human mind.

2.2 Visual Technologies and Sensory Case Studies

In the exploration of visual technologies and sensory case studies, the complexity of how our brains process visual information shall be foregrounded. While the human eye is responsible for registering up to 80% of all sensory impressions through sight, the interpretation and understanding of this visual data involves intricate neural mechanisms. In fact, there has been much more research on vision than any other sensory modality (Hutmacher, 2019). This section investigates how artists have harnessed visual elements to convey messages that resonate with diverse neurotypes and experiences. These artistic endeavors transcend the restrictive bounds of the sighted world, offering insights into how visual art can communicate in ways that are both universally accessible and deeply personal. The discussion will illuminate the creative ways in which artists have utilized visual technologies, not just for aesthetic purposes, but as tools to express and navigate the varied landscapes of human perception and cognitive experience.

In proceeding into the realm of visual technologies and their sensory implications, a recent study by the University of Waterloo (2023), offers crucial insights. The discoveries underscore the profound impact of this cognitive bias toward symbols on the processing of visual stimuli, implications that are critical for understanding the artistic examples discussed in the subsequent section. The study demonstrates that symbols are more memorable than words, providing a more effective anchor for abstract concepts. The enhanced recall of symbols is due to their unique and unambiguous nature, which renders them readily identifiable and less susceptible to diverse interpretations than words (Clark, 2006). The findings align with the *dual-coding theory*, suggesting that symbols, being coded both visually and verbally in the brain, have an enhanced memorability factor (Damayanti et al., 2023). Symbols or pictorial representations can elicit a stronger neurological response than words because they are processed faster and are more directly linked to the sensory and perceptual experience (Borghesani & Piazza, 2017). The brain regions involved in visual and spatial processing, such as the fusiform gyrus and the right hemisphere, are particularly activated by images and symbols, providing a more robust memory trace (Ganis & Kutas, 2003).

Along with neurotypical considerations of symbolic preference are those with various physical limitations. For instance, augmentative and alternative communication (AAC) systems, such as those utilizing pictograms, play a crucial role in aiding individuals with speech disabilities. These systems are particularly beneficial for individuals with autism, offering a means of communication that bypasses traditional language barriers. Pictogram-based applications, like PictoEditor, incorporate predictive functionalities that enhance the availability of desired pictograms for users. This approach becomes more effective when predictions are based on the frequency of pictogram use (Hervás et al., 2020). Additionally, Blissymbols, known for their minimalistic and semantically combinatory nature, are instrumental in tactile communication for AAC needs. Their design is geared toward maximizing tactile processing, thereby aiding in the representation of new words and meanings in diverse environments (Bliss et al., 2012). Furthermore, systems like

Pictogrammar, which link to the Simple Upper Ontology and PictOntology, utilize formal semantics to manage pictogram sets and generate natural language utterances from a sequence of symbols, dynamically adapting the communication board based on syntactic and semantic content (Martínez-Santiago et al., 2020). These advancements in AAC technologies exemplify how specialized communication systems can significantly improve the quality of life for those with speech and communication challenges. This preference for symbols over words has significant implications in the field of neuroarts, especially considering individuals with varying sensory processing capabilities, such as synesthesia or visual limitations. For instance, the use of VR technology to create immersive environments for people with synesthesia allows for a more intuitive and symbol-centric exploration of their unique sensory experiences (Lee, 2023).

Additionally, the study sheds light on the concept of lingual vision and the use of reflective surfaces by individuals who are deaf or hard of hearing, illustrating how visual technologies can be adapted to different sensory needs (Holmström, 2022). In the context of visual impairment, understanding the role of photoreceptor function, the optic nerve, and the occipital lobe is crucial in recognizing and appreciating art, paving the way for technology-enhanced sensory experiences (Fig. 2.3) (Othman et al., 2022). In other words, when it comes to visual impairment, it is important to understand how certain parts of the eye and brain work. This includes the photoreceptors (which are like tiny cameras in our eyes), the optic nerve (the cable that sends pictures from our eyes to our brain), and the occipital lobe (a part of the brain that helps us understand what we are seeing). Understanding these can help people with visual impairments experience and appreciate art in new ways, especially with the help of technology that enhances their sensory experiences.

Exploring various examples of technological intervention demonstrates the potential for multisensory integration. For instance, the exhibition *Sensorium* at the Tate Britain (2015), revolutionized art appreciation by offering a multisensory experience. The exhibit, made up of four rooms, featured four paintings from the Tate collection by Richard Hamilton (1922–2011), John Latham (1921–2006), David Bomberg (1890–1957), and Francis Bacon (1909–1992). Wearing wristband monitors, visitors could not only see but also hear, smell, taste, and touch the artworks through carefully curated sensory elements. Collaborating

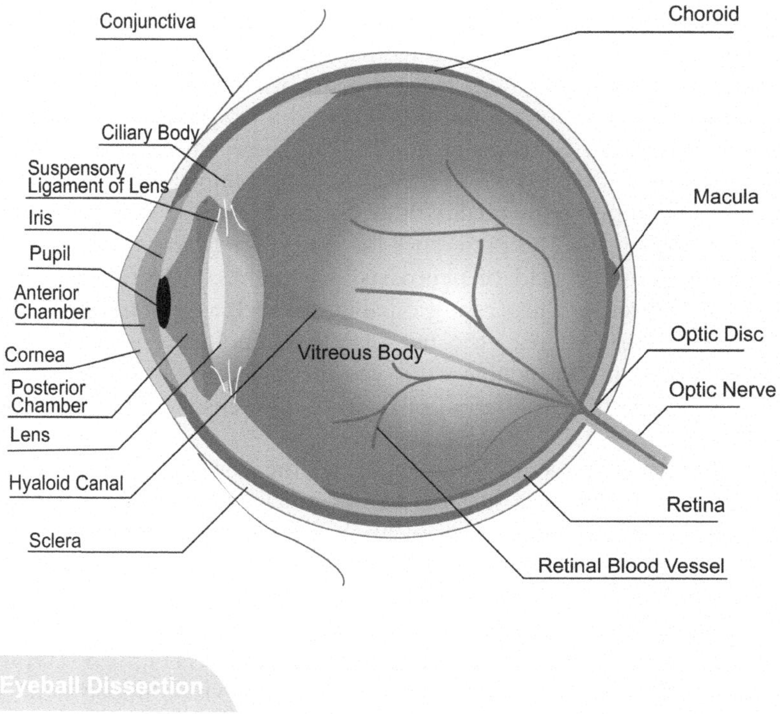

Fig. 2.3 Eyeball dissection, 2017 (*Source* From Wikimedia Commons, licensed under CC0)

with experts in sound, taste, scent, and touch, this exhibition transformed art into a holistic encounter. Visitors could, for example, listen to sounds corresponding to the theme of the work, smell scents inspired by the paintings, taste flavors evoked by the art, and feel physical forms echoing their essence. This innovative approach was realized through the partnership with Flying Object, a London-based creative agency that leveraged technology to tailor sensory experiences to individual works of art. *Sensorium* challenged traditional modes of art appreciation, providing an immersive journey that engaged multiple senses and redefined the art-gallery experience (Blundell, 2023). Such installations utilize advancements in visual technology to deepen the engagement with art, making it more accessible and resonant with a wider audience.

While multisensory experiences, like the *Sensorium* exhibition, demonstrate how visual experiences can be augmented with other senses, technology can also be adeptly employed to accommodate and enhance the artistic experience for those with various visual conditions, such as colorblindness. It should be noted that there are levels of colorblindness as illustrated in Fig. 2.4. New technologies present promising solutions for individuals with colorblindness, enhancing their ability to fully experience visual art. Applications leveraging machine learning (ML) algorithms have the proven potential to transmute colors into perceivable shades or patterns for those affected (Zhang & Tan, 2023). For instance, Color Oracle (Design for the Color Impaired) (https://colororacle.org/) is an application that interprets colors into distinct shades of gray, aiding in differentiation. The color blindness simulator for Windows shows in real time what those with common color vision impairments see to help adjust designs (Engeset et al., 2022). Another application, Chromatic Vision Simulator by Kazurnori Asada (1948–) (https://apps.apple.com/us/app/chromatic-vision-simulator/id389310222), provides simulations of various types of colorblindness, which could be invaluable in art exhibitions, enabling colorblind visitors to engage with artworks in a more meaningful manner. Beyond these, ColorADD (https://www.coloradd.net/en/) presents an innovative color-coding system that employs geometric shapes to signify different hues, enhancing accessibility within artistic spaces. Furthermore, Vischeck (https://www.vischeck.com/) offers a digital solution by adjusting color presentations on screens to accommodate the visual limitations of colorblind users and simulates colorblind vision. These technological interventions represent a concerted effort to ensure that visual art remains inclusive and enriching for all audiences, regardless of their visual capacities.

Continuing to explore technological solutions for various visual impairments, the focus shifts to the notable yet often overlooked effects of albinism on vision and the innovative methods technology provides for support. Albinism is characterized by reduced melanin production and results in several vision-related challenges due to the lack of pigment in the eyes. This leads to conditions such as light sensitivity, nystagmus, and photophobia, which can significantly affect visual acuity and overall sensory experience of individuals (Moreno-Artero et al., 2021). Neuroscientific studies have further indicated that albinism may also alter the structure of visual pathways, impacting the processing of visual inputs (Ather et al., 2019; Hoffmann & Dumoulin, 2015). For example, the

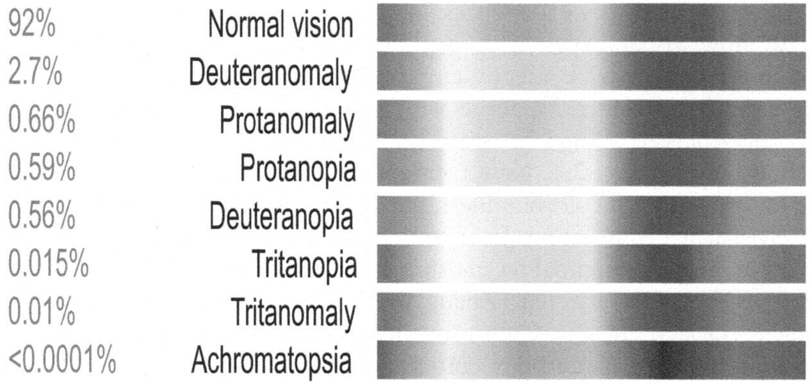

Fig. 2.4 Colorblindness, 2022 (*Source* From Wikimedia Commons, licensed under CC0)

development of the fovea, which is crucial for sharp vision may be affected and lead to difficulties in visual acuity and contrast sensitivity. Furthermore, there might be distinctive patterns in how the brains of individuals with albinism process visual information, especially in the domains of color vision and motion detection (Oduntan et al., 2002).

In addressing the specific challenges associated with albinism, especially those related to vision, the application of advanced technologies such as AR and VR shows significant promise. These technologies can be tailored to create adaptive visual environments, compensating for the unique visual needs of individuals with albinism. For instance, AR and VR can be programmed to adjust light levels, color contrasts, and motion perception in a controlled manner, thereby reducing issues like light sensitivity and enhancing visual clarity (Fig. 2.5) (Hsiang et al., 2022). Furthermore, these technologies can adapt or simulate real-world environments which cater to particular visual impairments and create a more accessible and comfortable visual experience. This adaptation significantly aids in routine activities and enhances the appreciation of visual arts, promoting a more inclusive and immersive experience for those with albinism. Concurrently, artists have employed similar technologies to immerse audiences in multisensory experiences.

In the realm where art intersects with technology, artists ingeniously utilize visual elements to convey messages that resonate across different

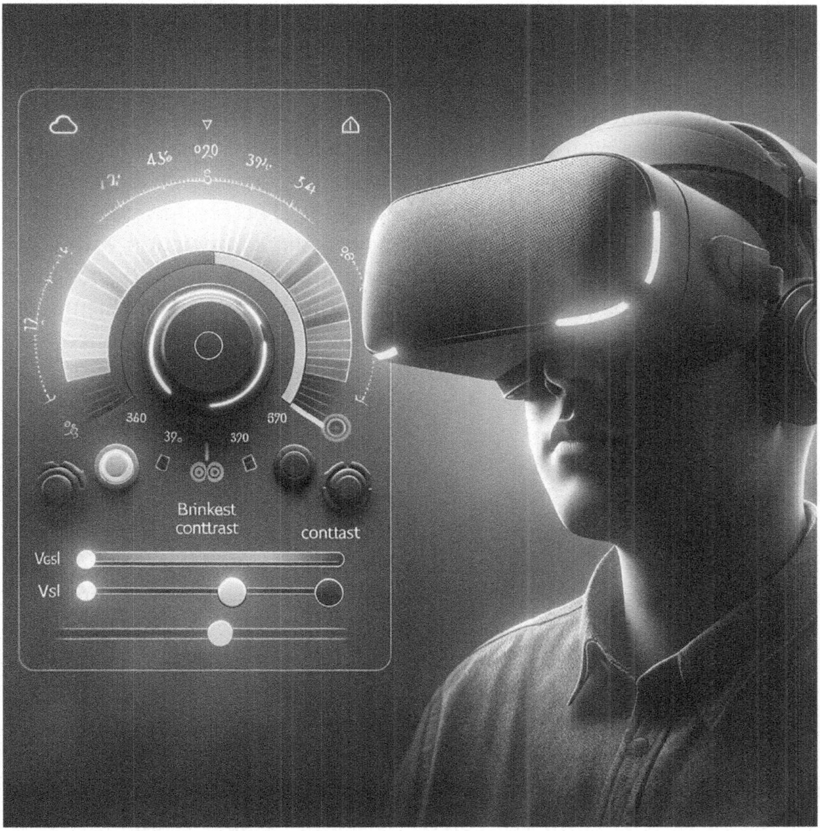

Fig. 2.5 VR adjustable light levels, 2023 (*Source* Permission of authors)

neurotypes and sensory experiences. The exploration has resulted in a diverse array of innovative and stimulating installations. In addressing visual impairment, over 80% of those affected can differentiate between light and dark (Müller et al., 2022). This fact is crucial for artists and technologists aiming to create inclusive art experiences. Leveraging what is known about photoreceptors and the processing of visual signals in the occipital lobe, artists can create works accessible and engaging for visually impaired audiences. This approach is evident in installations which use light and shadow to create immersive experiences, making visual art more inclusive and resonant with a wider audience.

Starting with the interplay of colors and soundscapes, *The Color of Rain* exhibit at the Museum of Arts and Sciences (MOAS) stands as a poignant exploration of color and perception, intertwining the elements of rain and artistic interpretation. Launched on March 26, 2022, the exhibit challenges visitors to reimagine the relationship between rain and color. A diverse group of 44 artists rose to this challenge, each contributing their unique vision to the exhibit. The resulting display is a symphony of umbrellas, each a canvas where rain's interaction with color is reimagined and expressed. Visitors are immersed in an ambiance filled with the sounds of rain, courtesy of CitiSound, further enhancing the sensory experience. As they move through the exhibit, they encounter a multitude of interpretations, each umbrella revealing a distinct story of how rain transforms and interacts with color. Tobey, curator of the exhibition, offers an insightful perspective on the inherent lack of color of rain and how that sets the stage for a transformative artistic journey. "Rain doesn't really have a color," Tobey noted, "But what I challenged the artists to do was to give their interpretation on their umbrella canvas of what rain looks like when it touches color" (Almenas, 2022). The intriguing concept posits rain as a magnifying glass, one that intensifies and deepens colors, lending them a newfound vibrancy and depth. Thus, the phenomenon becomes a central theme of the exhibit as the rain plays as catalyst to illuminate hidden perceptions of color.

Other artists have put color and the construction of space centerstage. In experiences like *Perfectly Clear* and *Hind Sight* (https://massmoca.org/event/james-turrell/) by artist James Turrell (1943–), illumination becomes a discrete, physical object. The artist employs light to sculpt space and manipulate perception in spaces that use carefully calibrated light to create the illusion of infinite space, offering a unique sensory experience that challenges the perception of depth and boundaries of the viewer. In his works, segments of the sky appear as if magically hovering within ceilings or walls; conventional architecture seems to dissolve; and vivid, geometric forms appear to defy gravity. The effects of changing light and space can also be seen in the 2013 exhibition Aten Reign at the Guggenheim Museum in New York. Figure 2.6 represents two separate photographs of the same spot in the ceiling to illustrate the change brought about by light and color. Initiated in 1966, Turrell's journey with light as a medium began with painting the windows of his Santa Monica studio, effectively obscuring natural light, while he explored the possibilities of light projections (Fallon, 2014). His art has consistently transformed through the creative modification of architectural elements, thus reimagining and redefining viewer interaction within their environment. The strategic use of light modifies the physical space and elicits a deep emotional reaction, encouraging reflection and introspection.

Further exploration into the convergence of art and technology involves adapting medical devices for artistic creation. For instance, EEG art is a project that delves into the realm of human emotions and their representation through digital art, a frontier where the human mind and technology converge in creative harmony (Barrera, 2021). EEG, or electroencephalogram, serves as the cornerstone of this artistic venture. The neurological technology captures the electrical activity of the brain, translating these signals into graphical data. The project, led by Random Quark, a creative studio managed by Theodorous Papatheodoru and Tom Chambers, employs EEG headsets to scan brain activity of participants. Individuals are immersed in a tranquil environment and asked to recall emotionally charged memories. The EEG data, representing the electrical

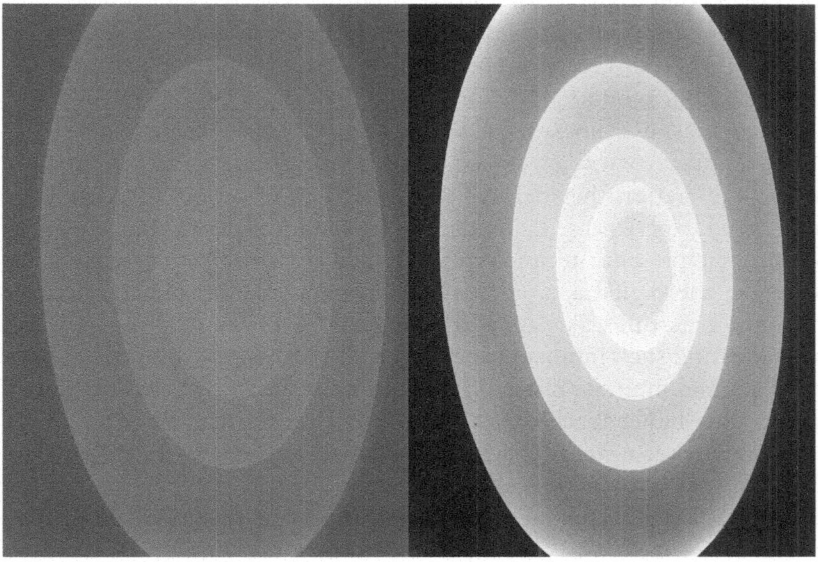

Fig. 2.6 James Turrell, *Aten Reign*, Gugenheim Museum, New York, 2013 (*Source* From Wikimedia Commons, licensed under CC-BY 3.0)

activity of the brain, is then used to create visual representations of these emotions.

The heart of this endeavor lies in the Geneva Emotions Wheel (Fig. 2.7), a tool which plots human emotions on a valence-activation graph. The approach of Random Quark filters these emotions, focusing on seven major ones: joy, sadness, anger, love, disgust, fear, and surprise. Each emotion is given a unique color, creating a vivid and dynamic portrayal of the participant's emotional state. Notable case studies from this project include the Saatchi & Saatchi Wellness Emoscape, which sought to combine scientific precision with artistic flair. Participants reminisced about significant life events, and their brainwaves were transformed into complex digital paintings, each a unique representation of the associated emotions. These paintings, exhibited at the Truman Brewery gallery in East London, showcased the intricate and personalized nature of EEG art. This project is not only a testament to the artistic potential of neurotechnology but also a reflection on the future of human–computer interaction. As Tom Chambers envisions, the integration of emotional context in computing could lead to more adaptive and empathetic technological interactions. The possibilities are immense, from creating art that captures our deepest emotions to designing environments that respond to our mood.

Along the same lines as EEG art, is the EEG KISS project (2014–2022) (https://www.lancelmaat.nl/work/e.e.g-kiss/), an avant-garde artistic exploration that delves into the realms of intimacy, technology, and shared experiences. Experienced in exhibitions from Amsterdam to Beijing, tele-presence technologies are used to scrutinize the concept of kissing in the digital age, addressing questions around the transference of intimate gestures like a kiss into the online realm and the feasibility of quantifying such intimate experiences. At the core of this project once again lies the innovative use of EEG technology to translate the act of kissing into biofeedback data. The artists deconstruct the physical act of kissing to reconstruct a new, synesthetic ritual of shared experiences in what Joyce Roodnat called "Another kind of Rodin," as though the experience was a contemporary take on the Kiss by the artist (Fig. 2.8).

The performance installation, *Digital Synaesthetic EEG KISS*, stands as a testament to this artistic inquiry. Participants are invited to engage in a live kissing experiment while adorned with EEG headsets, their brainwaves measured and manifested as EEG data. This data is then used to create an immersive datascape surrounding the kissers, integrating them

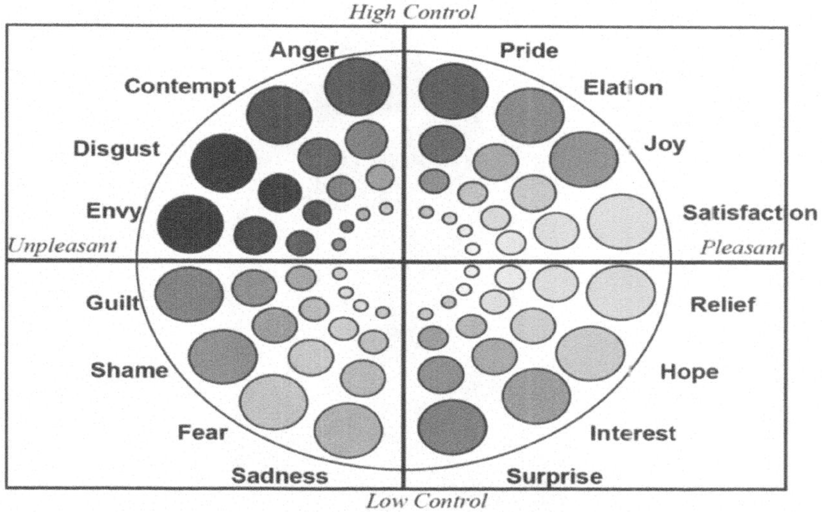

Fig. 2.7 Geneva emotion wheel (*Source* From Sacharin et al. [2012])

within a floor projection of streaming EEG data. Accompanying this visual spectacle is a soundscape generated from the EEG data, effectively translating the brainwaves of the kissing individuals into a musical score. The audience becomes an integral part of this experience, enveloped in the sound and visual representation of the kiss. As such, the project goes beyond mere technological demonstration and serves as a platform for participants to attribute personal meaning to the abstract and often mystifying EEG data visualizations of their kisses. Importantly, the project avoids scientific interpretation and validation, instead inviting individuals to interpret the data based on their shared memories and imagination. The result is often seen as intimate co-creation, with participants viewing the data as a "Portrait of our kiss." Furthermore, each unique EEG KISS visual data sequence is preserved in a database, available to be printed as an EEG KISS Portrait, while the corresponding soundscape is saved for public sharing. This facet of the project prompts crucial contemplation on intimacy and big data, challenging the safeguarding of personal data against the formation of new hybrid, relational rituals which permit public intimacy.

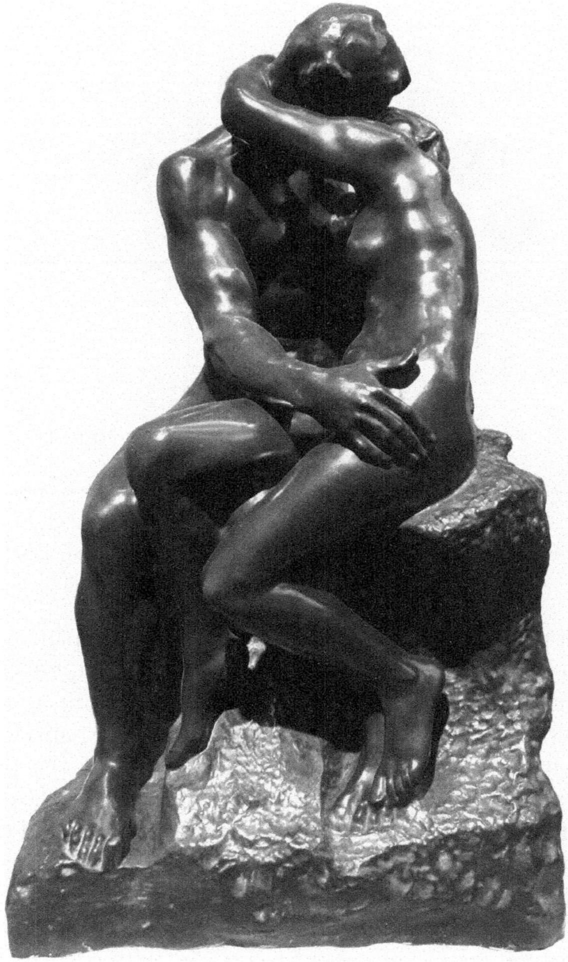

Fig. 2.8 Auguste Rodin, *The Kiss,* between 1898 and 1918. Bronze (*Source* From Wikimedia Commons, licensed under CC0)

As we transition from exploring visual technologies to auditory experiences, the role of language in shaping perception becomes increasingly evident. Coyote (https://coyote.pics/), an innovative platform providing visual descriptions, serves as a bridge between the visual and auditory realms, particularly enhancing art experiences for individuals with visual impairments. This collaboration with the Museum of Contemporary Art in Chicago underscores the potential for auditory descriptions to enrich the appreciation of visual arts (Lupton & Lipps, 2018). Such initiatives highlight the symbiotic relationship between visual and auditory senses, emphasizing that perception is not isolated within a single sensory modality but is profoundly influenced by language and cultural context. As we delve into auditory experiences, we carry forward the understanding of how language not only differentiates colors but also structures the sensory world, driving further investigation into the dynamic interplay between visual art and auditory narrative. This exploration encourages us to consider the broader spectrum of neuroarts, where the fusion of sight and sound, influenced by linguistic nuances, creates a more inclusive and deeply resonant artistic experience.

2.3 Auditory Technologies and Sensory Case Studies

Although balance, spatial awareness, and proprioception are essential in human sensory experience, discourse on art primarily centers on visual perception, frequently overlooking the auditory dimension. Yet, auditory processing is crucial in our sensory engagement, providing vital spatial information about our surroundings, second only to vision. Notably, music and sound constitute the most extensively researched area in neuroarts. Therefore, it is essential to recognize and emphasize sound as more than a secondary element in sensory experiences, acknowledging its profound impact on perception and interpretation of the world around us. Furthermore, the phenomenon of synesthesia, where sensory experiences cross over, is a notable aspect in the arts with auditory-visual synesthesia being one of the most common manifestations (Goller et al., 2009). Many artists, as noted, including Kandinsky, and contemporary musicians like Pharrell Williams (1973–), have experienced synesthesia, where auditory stimuli trigger visual perceptions, and vice versa. Williams, for instance, perceives music in terms of colors, a sensory crossover that significantly influences his songwriting process (Harkness et al., 2023). Kandinsky similarly experienced synesthetic connections, reportedly being able to "hear" the colors in his vibrant paintings (Cardullo, 2017). This interplay of senses, particularly between sound and sight, enriches the creative process and underscores the multisensory nature of artistic expression.

Beyond multisensory integration, the significance of auditory perception in human life is evident in the extensive array of technologies developed to enhance hearing or mitigate hearing loss. These innovations include hearing aids, cochlear implants, bone-anchored hearing systems, auditory brainstem implants, and middle ear implants. Additionally, there are sound amplification devices, frequency modulation (FM) systems, advanced digital hearing aid technology, tinnitus maskers, sound therapy systems, and directional microphones. Further aiding auditory communication are real-time captioning devices, speech-to-text software, loop systems for telecoil-equipped hearing aids, sound field systems, vibrotactile devices, personal amplifiers, smartphone hearing aid applications, hearing aid compatible phones, adaptive alarm systems, and visual as well as vibrating alert systems (Kim & Kim, 2014; Tharpe et al., 2008). Each of these technologies addresses various aspects of hearing enhancement or compensation, reflecting the critical role of hearing in human interaction and well-being.

These diverse auditory technology devices offer insights into the interaction between sound and the brain, illustrating the complexity of auditory processing once sound enters the ear. Initially, sound waves are captured by the outer ear and funneled down the ear canal, striking the eardrum, which causes the eardrum to vibrate, a movement that is then transmitted through the small bones of the middle ear known as the ossicles (Fig. 2.9). These vibrations are further conveyed into the cochlea, a fluid-filled structure in the inner ear, inside of which tiny hair cells are stimulated by the fluid's motion, converting these mechanical vibrations into electrical signals. These electrical impulses in turn are then transmitted via the auditory nerve to the brain, specifically to the auditory cortex. In the auditory cortex, these signals are processed and interpreted, allowing us to perceive and understand the sounds as distinct entities, such as speech, music, or environmental noises. Through the translation of sound waves into meaningful information by the brain, different technologies such as hearing aids and cochlear implants have been pivotal in aiding those with hearing impairments. Hearing aids amplify sound to make it more accessible to the user, whereas cochlear implants bypass damaged parts of the ear and directly stimulate the auditory nerve, offering an alternative pathway for sound perception (Green et al., 2022).

The multitude of innovations designed to assist hearing underscores the critical role of sound in daily life, mobility, communication, and

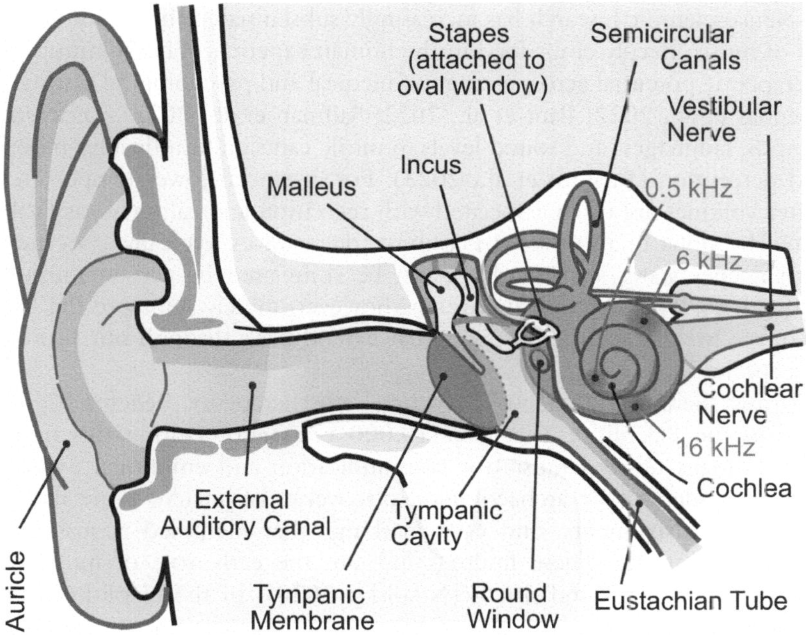

Fig. 2.9 Anatomy of human ear with cochlear frequency mapping (*Source* From Wikimedia Commons, licensed under CC-BY 2.0)

enjoyment. These advancements reflect the scientific understanding of auditory perception as essential for navigating environments, interacting socially, and experiencing pleasure (Zhao et al., 2018). From the vibration of the eardrum to the intricate workings of the auditory cortex, the profound effects that music and sound can have on the human brain and emotions are historically undeniable. The ability of music to evoke strong emotional responses, influence cognitive processes, and facilitate social interaction makes it a vital element in the study of neuroarts and multisensory experiences. Neuroarts seeks to understand and bridge these sensory differences, offering alternative pathways for emotional expression and reception (Guga & Uspenski, 2016). By moving beyond a visual-centric approach and embracing the auditory domain, a more comprehensive understanding of sensory experiences and their impact on human cognition and emotion can be achieved.

Neuroscientific research has increasingly substantiated the health benefits of music, deeply embedded in the human experience, highlighting its therapeutic potential across a range of medical and psychological settings. (Cephas et al., 2022; Pant et al., 2022; Saifman et al., 2023). Different tempos, languages, and sound levels in music can significantly alter mood and perception (Imasato et al., 2023). For instance, slower tempos and softer volumes are often associated with relaxation and calmness, as facilitated by alpha brain waves. In contrast, delta waves, typically associated with deeper states such as sleep, can be stimulated by certain musical frequencies. These recent discoveries, however, merely reinforce the use of music for therapeutic purposes that extend back through our history as a species.

Music therapy is an age-old practice with therapeutic benefits beginning in the Paleolithic Era, over 10,000 years ago. During this time, early humans utilized music for communication and emotional expression, as evidenced by archaeological discoveries of ancient bone flutes, percussion instruments, and cave markings that identified acoustically resonant locations. These findings indicate the early role of music in communal activities and rituals (Nikolsky, 2020). In the Neolithic Era, music saw significant evolution within permanent settlements globally. Various instruments like harps and complex percussion tools were developed, and rudimentary music notation began to appear, as seen in clay tablets from ancient Mesopotamia. The period also in fact marked the growing significance of music in religious ceremonies and social gatherings (Pomberger et al., 2021).

The therapeutic power of music was also recognized by ancient Greek philosophers. For instance, Plato extolled music as a source of joy and healing, famously stating that it gives "soul to the universe, wings to the mind, flight to the imagination" (Dhar & Das, 2020). The pupil of Plato, Aristotle, understanding the profound impact it had on human character, also advocated for its inclusion in the education of the young. Throughout history, different cultures have acknowledged the healing powers of music (Provenza, 2020). Ancient Egyptians incorporated it into religious ceremonies (Fig. 2.10) (Teeter, 2011), Native American tribes like the Navajo used music and dance in healing rituals (Frisbie, 1980), and traditional Chinese medicine employed specific musical tones and rhythms to balance the body's energy (*qi*) (Tao et al., 2016). In the Middle Ages and Renaissance, the Christian Church played a crucial role in popularizing music among the masses. Congregational hymn singing during church services became a powerful medium for religious devotion, teaching, and therapeutic experience, especially for the non-literate populace (Fenlon, 2002; Lord, 2008; Monti, 2012).

Following the early modern period, the rise of the Industrial Revolution and Age of Enlightenment in the eighteenth century witnessed the parallel development of studies on the human nervous system and music therapy (León-Sanz, 2016). American physician Benjamin Rush (1745–1813) recognized the potential of music to improve mental health, while his student, Samuel Mathews, conducted experiments exploring the effects on the nervous system, laying the groundwork for modern music therapy (Heller, 1987). As seen in their Tranquilizing Chair (Fig. 2.11), such early experiments included having a patient sit in a chair, immobilized with straps at the shoulders, waist, arms, and feet and have a box-like apparatus confining the head to focus the sounds. With these early rejections of mental illness as the result of demonic possession, new forms of psychiatric and psychological treatment arose. For instance, the psychologist Everett Thayer Gaston (1901–1970), often referred to as the

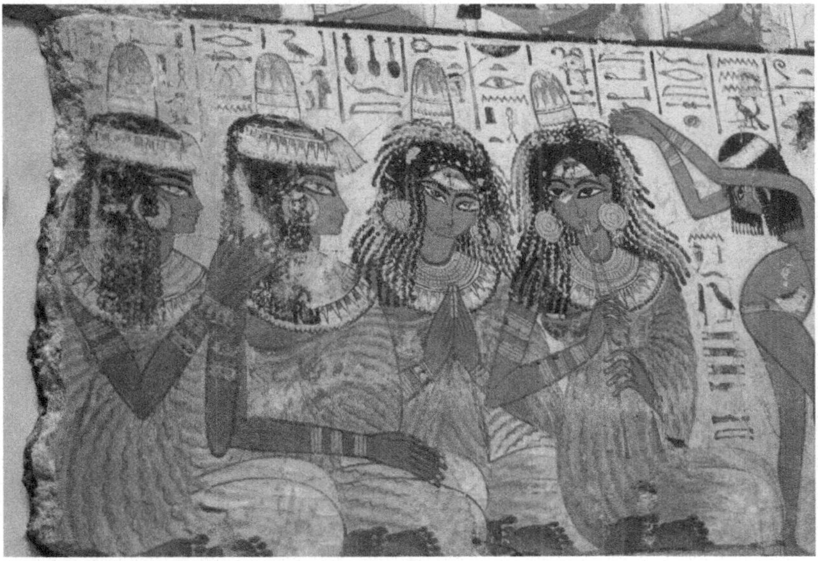

Fig. 2.10 Tomb of Nebamun, 1350 BCE, Thebes (Luxor) (*Source* From Wikimedia Commons, licensed under CC-BY 3.0)

"father of music therapy," was instrumental in establishing it as a recognized discipline in the United States in the 1940s–1960s, while the British music therapist Mary Priestley (1925–2017) contributed significantly to the development and respect of the field, developing Analytical Music Therapy, a synthesis of psychoanalytic theory and music therapy (Bunt & Stige, 2014).

Since the acceptance of music therapy as a field of its own, several phenomena have been observed. The "Mozart effect" is one such example that stands out as a compelling inquiry into the influence of music on cognitive function. Originating from a 1993 study by Frances Rauscher and Hinton (2006), the concept suggests that listening to the classical music of Wolfgang Amadeus Mozart (1756–1791) may improve spatial reasoning skills. Despite mixed evidence, the research has deepened our interest and understanding of the neurological impact of music (Shi, 2020). Follow-up studies using other classical composers, such as

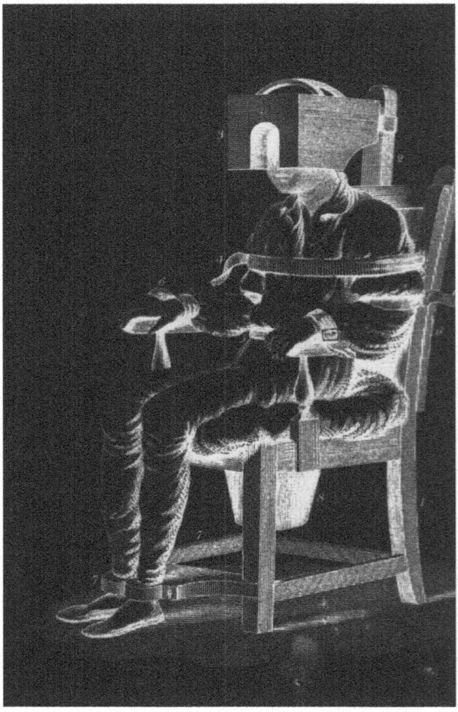

Fig. 2.11 Tranquilizing Chair of Benjamin Rush, 1811. National Library of Medicine (*Source* From Wikimedia Commons, licensed under CC0)

Ludwig van Beethoven (1770–1827) and various types of rock music have utilized EEG to measure attention and dopamine release, illustrating that classical music, in particular, can enhance cognitive processing without the influence of musical preference. This was evidenced by the accelerated processing of visual stimuli under the influence of classical music (Santhosh et al., 2020; Tai & Kuo, 2019). Moreover, personal enjoyment of music was linked to increased alertness and mood, akin to the effects of other pleasurable experiences, suggesting the artform's broader capacity to enhance mental well-being (Hennessy et al., 2021).

The ability of music to engage the brain encompasses a complex interplay of auditory signals and neurochemical reactions, notably the release of dopamine, which is associated with inducing feelings of pleasure (Ferreri et al., 2019). Contrary to previous beliefs, studies have indicated that even somber tones, such as those from Antonio Vivaldi's (1678–1741) concerto *L'inverno* (Winter) from the *Four Seasons* (1718–1720), can induce strong dopamine responses (Riby et al., 2023). While activities such as eating and sex have clear evolutionary purposes, the benefits of music, although less direct, are significant for social bonding and cultural continuity (McGrath & Brennan, 2020). The nucleus accumbens, central to the reward system of the brain, activates the dopamine pathway when we interact with music, amplifying positive emotions (Gold et al., 2019). Furthermore, the amygdala, the center for emotional response, is notably responsive to melodic stimuli (Fraile et al., 2023). EEG tests confirm that music promotes neural communication, shaping our emotional and cognitive experiences: happy tunes generally foster positive, outward-looking emotions, whereas sad music prompts reflective, inward-focused thoughts (Ueno & Shimada, 2023). These insights hold promise for music therapy, particularly for those grappling with emotional or neurological challenges, as music offers a means for emotional catharsis and introspection. The power of music to trigger potent autobiographical recollections can lead to transformative experiences, such as the deeply moving response of the ex-ballerina Martha González Saldaña when she hears Pyotr Illyich Tchaikovsky's *Swan Lake* (1875–1876). The 2019 recorded session (https://youtu.be/owb1uWDg3QM) where the former prima ballerina, who suffered from dementia and memory loss, had the melody played for her had an immediate reaction and began swaying as though dancing (Tsioulcas, 2020).

These instances bear witness to the potent responses music can elicit in the brain, prompting ongoing research into how different musical rhythms and compositions affect emotional health and neurochemical dynamics, as well as how artists might utilize these sensory experiences. In the artistic domain, technological advancements are reshaping and broadening the spectrum of sensory experiences. For instance, the American deaf artist Christine Sun Kim (1980–) uses her work to explore sensory perceptions. Based in Berlin, the sound artist incorporates sign language, text, and sound to reflect on how deaf people interact with their surroundings. For her project *One Week of Lullabies for Roux* (2018), Kim orchestrated a collaboration with seven parent-friends to create unique, calming soundscapes aimed at soothing her newborn, Roux, into slumber. Her directive for these lullabies was to avoid lyrics and focus on low frequencies, ensuring she could both monitor and feel at ease with the sounds introduced into her baby's auditory environment. This project, symbolized by an artistically crafted bench resembling a color-coded pillbox, metaphorically represents these soundtracks as a form of daily nourishment (Cocozza, 2022).

Similarly, the Mexican-Canadian electronic artist Rafael Lozano-Hemmer (1967–) brings together art, design, and technology in his interactive exhibitions such as *Pulse* (2018–2019) at the Hirshorn Museum. The artist transformed the Second Level of the museum with immersive installations that utilized heart-rate sensors to create kinetic and audiovisual experiences based on biometric data of visitors. These installations captured and visualized biometric signatures as sequences of lights, soundscapes, and animated fingerprints, integrating them into a rhythmic, communal archive. The exhibit showcased three major works: *Pulse Index*, which recorded fingerprints and heart rates on a large-scale projection grid; *Pulse Tank*, where the pulses of visitors created ripple patterns in illuminated water tanks; and *Pulse Room*, featuring hundreds of light bulbs pulsating with past the heartbeats of visitors. Visitors interacted with these installations, contributing their biometric data, thereby blending anonymity with a sense of community (Lozano-Hemmer, 2010).

Introducing a unique blend of technology and sound, sound artist Yuri Suzuki's (1980–) *Sonic Bloom* installation (https://yurisuzuki.com/projects/sonic-bloom) presents a site-specific auditory experience that captures and melds themes of people, nature, and the environment. The installation fosters human interaction within spaces traditionally void of such engagements. By inviting participants to not only communicate with each other but also to connect with their environment, the sound experience transcends conventional boundaries of art and communication.

Its digital counterpart extends the reach of this experience, allowing a global audience to share in the enchantment of the physical installation. In this digital space, voice recordings are artistically transformed into flower animations and then whimsically "planted" onto a virtual map of Mayfair (where the installation is located) creating an interactive and immersive exploration of universal communication (Bode, 2022).

Auditory experiences continue to evolve, propelled by technological advances, leading to ever greater personalization. As an example, the "cyborg artist" Neil Harbisson (1982–) takes sensory integration a step further with an antenna implant in his skull (Fig. 2.12) that translates colors into sounds, offering a synesthetic experience and expanding the boundaries of perception. Born completely colorblind, the prosthetic device that he refers to as an "eyeborg" allows him to hear colors as interpreted by the device as audible frequencies. The device allows the artist to experience a broader spectrum of colors, including ultraviolet (Harbisson, 2019). Less invasive auditory augmentation devices also exist for the general public. Other advances in auditory technologies, like PerL's sonic adaptive headphones, offer personalized listening experiences, and apps like Coffitivity and Study Ambience, which simulate ambient environments to boost focus and creativity (Droumeva, 2021). These technological advancements underscore the evolving synergy between personalized sensory experiences and technological innovation, opening new avenues for artists.

Such personalized auditory experiences need not be isolated to the sense of hearing but can also coalesce with other stimuli for unique sensations. One notable crosspollination is between auditory explorations and the tactile dimension. The fusion of auditory and tactile experiences in art offers a unique sensory exploration, exemplified by the works of Christine Sun Kim and Cevdet Erek (1974–). As noted, Kim was deaf from birth, therefore, she delves into her personal relationship with sound. In the Greater New York exhibition at MoMA PS1, her piece *Game of Skill 2.0* (2015–2016) (https://vimeo.com/142659892) integrated sound and touch, engaging viewers to move a staff-like device along a velcro strip, which, in turn, altered the radio's sound levels and speeds. Similarly, Cevdet Erek, with a background in architecture and as a member of Nekropsi, ventures into the art world with interactive projects. His award-winning work *Shore Scene Soundtrack* (2012) (https://vimeo.com/2867660) invites participants to mimic the sound of the ocean by rubbing their hands on a carpet, merging auditory and tactile elements to

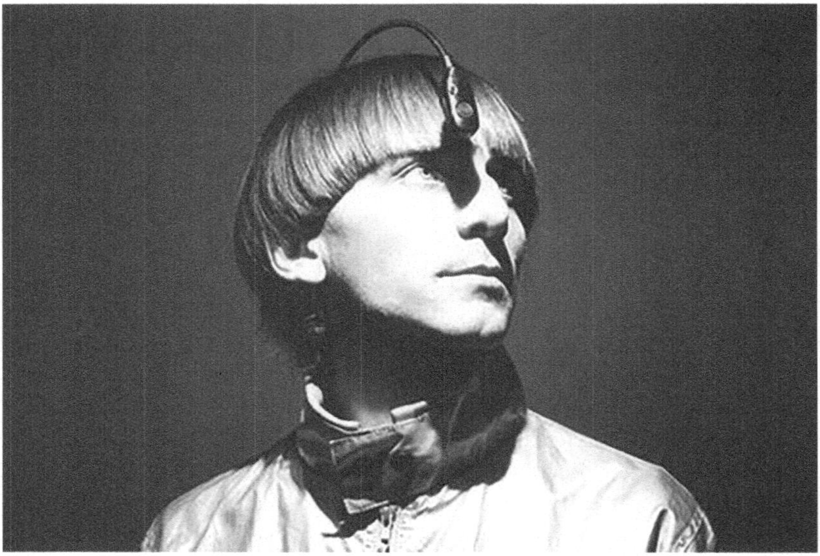

Fig. 2.12 Neil Harbisson, October 2012 (*Source* From Wikimedia Commons, licensed under CC-BY 2.0)

create a vivid seascape (Towers, 2019). These artists exemplify the creative potential of blending senses to enrich artistic experiences. Through these tactile ventures, neuroarts continues to illuminate the intricacies of human cognition, inviting us to experience and appreciate the complex beauty of sensory integration.

2.4 Tactile Technologies and Sensory Case Studies

Although tactile sensations, activated by touch receptors in the skin, do not process as high a percentage of environmental sensory stimuli as vision, they are crucial for comprehending the human condition and proprioception. Proprioception, often described as the "sixth sense," refers to the ability of the body to sense its position, movement, and spatial orientation in space and is a fundamental aspect of our sensory experience, though less overtly recognized compared to other senses like sight (Tuthill & Azim, 2018). Proprioception, a critical aspect of the somatosensory system, is the internal sense that tells us where our body parts are without having to look. It's mediated by proprioceptors, specialized sensory receptors located in muscles, tendons, and joint capsules, providing continuous feedback to the central nervous system about the position and movement of our body parts. For instance, even if our eyes are closed, we can sense if our arm is raised or bent, facilitating a nuanced awareness of our limbs' positioning and movement. This internal focus differentiates proprioception from externally oriented senses like touch, making it essential for coordinated movement, balance, and executing complex motor tasks (Zhan et al., 2023).

Simultaneously, the somatosensory system, encompassing skin, muscles, and joints, detects various stimuli such as light touch, deep pressure, pain, and temperature. It plays a dual role by contributing to both interoceptive and exteroceptive processing. This means it not only helps in understanding stimuli originating inside the body, thereby impacting our higher-order awareness of our physiological state but also aids in perceiving external stimuli, enhancing our immediate external awareness (Abraira & Ginty, 2013). These integrated sensory inputs are foundational for our interaction with the environment and self-regulation within it.

Therefore, while proprioception and touch are functionally distinct, they often work together to provide a comprehensive understanding of both the internal state of the body and its interaction with the external world. For instance, when holding an object, tactile receptors enable feeling its texture and shape, while proprioceptors allow us to adjust grip strength and arm position (Qi et al., 2023). The external journey of tactile perception thus begins at the skin, where a variety of touch receptors detect stimuli such as pressure, temperature, and vibration (Fig. 2.13). These receptors are connected to neurons that transmit signals through the spinal cord to the brain's thalamus, acting as a relay station. Once these signals reach the thalamus, they are then forwarded to the somatosensory cortex located in the parietal lobe. This region of the brain is crucial for interpreting tactile information and integrating it with other sensory data. The somatosensory cortex is adept at discerning where on the body a touch occurred and what kind of touch it was, whether a gentle caress or a sharp poke (Puts & Cascio, 2023). Moreover, tactile experiences often engage brain regions traditionally associated with visual processing (Calzavarini, 2023). The tactile sensory system intricately intertwines with the visual processing regions in the brain, illustrating a complex relationship between touch and sight. Touch receptors in our skin serve as the initial point of contact for tactile experiences, detecting various stimuli such as pressure, temperature, and texture. These receptors send signals through neurons in the spinal cord, which then relay the information to the thalamus. The thalamus acts as a crucial hub, distributing these tactile signals to the somatosensory cortex in the parietal lobe of the brain, where they are processed and interpreted. This overlap in sensory processing areas underscores the interconnected nature of our sensory systems and the capacity of the brain for integrating multisensory information.

Touch also plays a vital role in emotional and social interactions, largely due to the release of oxytocin. Often referred to as the "love hormone," oxytocin is released in response to physical touch and is associated with feelings of trust, bonding, and happiness (Morrison,

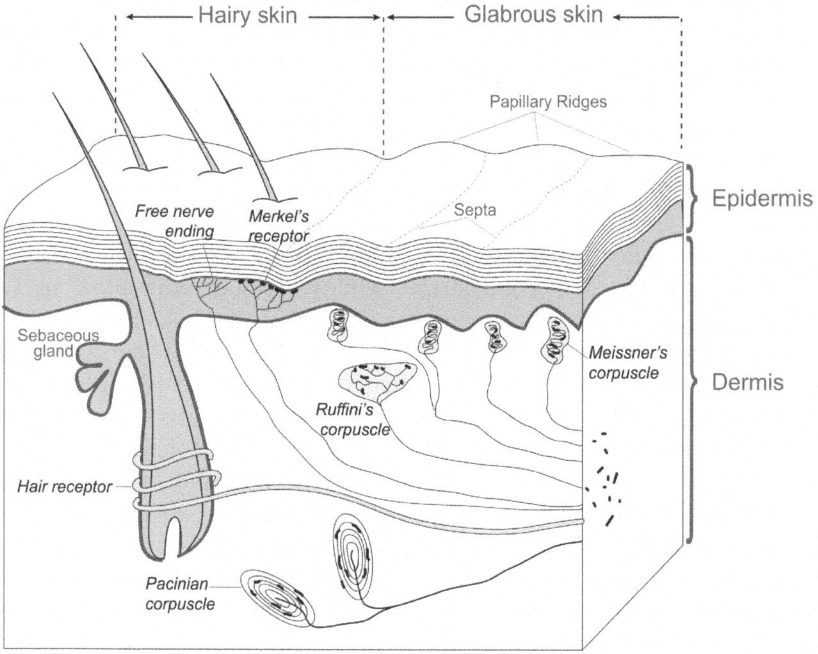

Fig. 2.13 Tactile receptors in the skin (*Source* From Wikimedia Commons, licensed under CC-BY 3.0)

2023). This neurochemical response enhances the emotional resonance of tactile experiences, further cementing touch as an integral component of human connection and well-being. The exploration of tactile technologies extends these understandings into the realm of sensory experience and artistic expression. By leveraging the complex neural networks involved in touch perception, artists and technologists create immersive experiences that not only stimulate the skin but also engage the brain in profound and emotionally resonant ways. This section will delve into the diverse applications and innovations in tactile technology, examining how they are utilized in sensory case studies to create novel and meaningful sensory experiences.

The exploration of tactile technologies in neuroarts is a journey through historical precedents and modern advancements, each contributing significantly to sensory experiences. This narrative begins with the foundational work in tactile communication, notably the development of Braille (Fig. 2.14), and advances to contemporary innovations in haptic technology and tactile headsets. In fact, the history of tactile communication is deeply intertwined with the story of Braille. Louis Braille (1809–1852), a 12-year-old student in Paris, revolutionized tactile reading for the visually impaired in 1821. Building upon an embossed typeface by Valentine Hauy (1745–1822) and a dot-and-dash system used for night-time army communication, Braille developed a tactile code based on six dots, creating symbols that were significantly easier to read than Roman letters (Thompson & Christian, 2022). The system, despite gaining traction more rapidly in Europe than in the United States, laid the groundwork for tactile communication. In the United States, Samuel Gridley Howe (1801–1876), directing the first school for the blind, created the Boston Line Letters in 1835. These tactile letters featured strong distinctions from each other, aiding in differentiation and recognition (Fulas, 2023). Following this, Elia Chepaitis in 1988 introduced a modified Latin alphabet for tactile use. This contemporary counterpart to Howe's system employed raised rectangular boxes with distinctive gaps and ticks, providing a clear tactile reference to the Latin alphabet (Chepaitis et al., 2004).

Advancements in recent years have seen a significant leap from these historical foundations. The Tactile Picture Book Project (https://www.colorado.edu/atlas/tactile-picture-books-project-build-better-book), spearheaded by Tom Yeh at the University of Colorado Boulder, represents a modern iteration of tactile reading. This project offers a digital

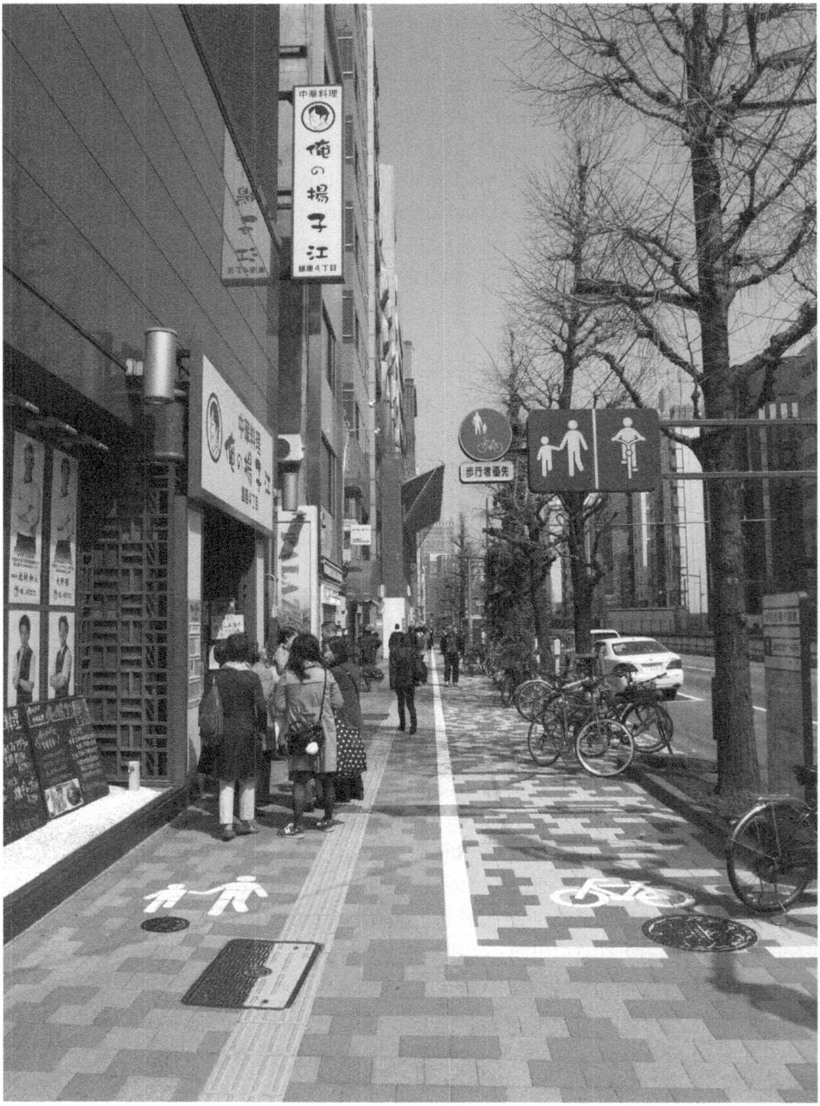

Fig. 2.14 Tactile path (*Source* From Wikimedia Commons, licensed under CC-BY 2.0)

library of books with 3D models representing various elements, allowing for an interactive tactile reading experience (Dalton & Musetti, 2018). But recent advances move beyond updates to traditional braille. Haptic feedback technology, for instance, which utilizes touch-based feedback like vibrations and pressure, emerged as a key innovation (Huang et al., 2022). An early example of this is in The "Museum of Pure Form," as conceptualized by Bergamasco et al. (2002), represents a pioneering integration of haptics technologies within the domain of cultural heritage, offering an immersive VR system that enables users to interact with digital models of 3D art forms and sculptures via both touch and sight. Developed at PERCRO in Pisa, Italy, this innovative system is being realized in two distinct formats: one designed for installation within various museums and art galleries across Europe, providing an enhanced experiential layer for visitors, and the other for implementation within a CAVE environment, offering a fully immersive virtual experience. While the technology is most well known in gaming with controllers providing haptic feedback to players, it is used in a number of fields now. Also, the field of sound-absorbing furniture intersects auditory and tactile experiences. These furniture designs incorporate materials that reduce ambient noise, creating a more comfortable environment for those sensitive to sound (Smardzewski et al., 2015).

Recent strides in tactile and haptic technologies mark a considerable evolution in sensory experiences, especially for those with visual impairments or unique sensory requirements. These innovations extend beyond conventional boundaries, offering new means of interaction and perception. A notable development is the placement of tactile replicas accessible to individuals using wheelchairs (Fig. 2.15). These replicas, often situated on turntables, allow visitors to feel and manipulate objects without needing to move their bodies (Eardley et al., 2016). The approach underscores the spatial aspect of touch, which relies on proprioception—the awareness of the position and movement of the body. Such awareness, as noted, is facilitated by receptors distributed throughout the skin, muscles, tendons, and joints, working in tandem with visual and auditory cues to contribute to a sense of movement and orientation. The concept of subjective skeletal space, as proposed by environmental psychologist James J. Gibson (1904–1979), is central to this experience, enabling a fleeting geometric schema of posture in space. The integration of body-based and skin-based touch leads to haptic perception, a system defined by Gibson where individuals feel an object relative to their body and vice versa (Gibson, 2002). Receptors in the muscles and skin relay signals about motion, pressure, pain, heat, and resistance, crafting a comprehensive sensory experience.

Urban design has also benefitted from tactile technological developments. For example, Tactile City (https://www.tactilecity.org/) is an experimental project that envisions a citywide tactile communication system for visually impaired pedestrians in New York City, New York. Conceived by Cooper Union students, it uses different textures to denote points of interest like bus stops and entrances (Phillips, 2015). The concept builds on the tactile paths (Fig. 2.13) invented in Japan by engineer and inventor Seiichi Miyake (1926–1982) in 1967 (Garofolo, 2022). Likewise, the accessibility work of Joshua Miele (1969–) should also be noted as a breakthrough for tactile maps (Fig. 2.16). His first-generation tactile maps utilize online geographic information systems to create user-centered, tactile maps for blind pedestrians. These maps, including San Francisco subway line maps, combine tactile features with audio explanations, accessed via a digital pen. Haptic signals vary in intensity and pattern to convey different messages. These signals can tap against the skin in precise patterns or blend into an overall texture, creating

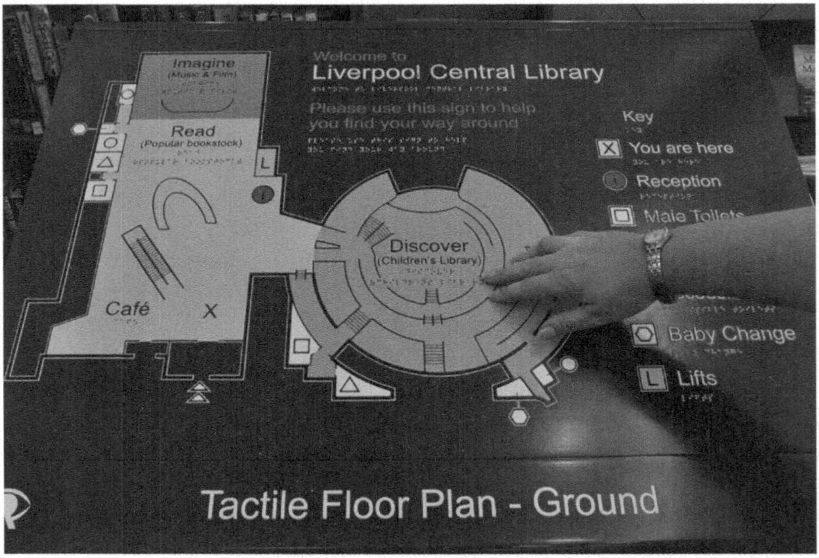

Fig. 2.15 Tactile floorplan of Liverpool Central Library (*Source* From Wikimedia Commons, licensed under CC-BY 2.0)

tangible illusions of objects and actions. Ultra haptic technologies, for instance, manipulate air pressure to simulate touch sensations, enhancing the perception of virtual objects (Morash et al., 2014).

Aside from navigating the physical environment through the use of tactile maps and paths, other systems like Feelipa (https://feelipa.com/) and Color ADD (https://www.coloradd.net/en/), both developed in Portugal, offer innovative solutions for translating colors into tactile forms. Feelipa uses basic geometric shapes to represent primary colors, while Color ADD employs a modular system based on parts of a square to depict primary and secondary colors, plus black and white. These systems can be applied to various products, aiding both color perception and reinforcing color knowledge. Notably, research indicates that color can significantly boost object recognition for individuals with low vision (Khan et al., 2012; Shukla & Verma, 2019). Furthermore, objects become more memorable when experienced through multiple senses, emphasizing the importance of multisensory integration in enhancing cognitive functions (Cox & Guillemin, 2018). In neuroarts, these tactile technologies are not standalone developments but are integrated into a larger sensory framework. They complement auditory and visual experiences, enriching the overall sensory journey. From historical tactile

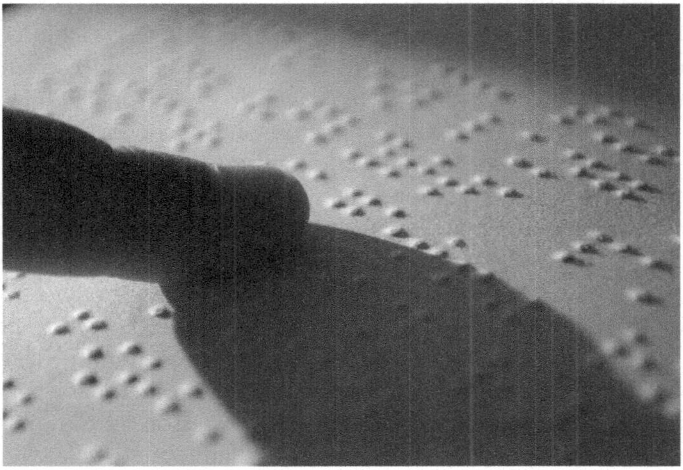

Fig. 2.16 Reading Braille (*Source* From Wikimedia Commons, licensed under CC-BY 2.0)

communication systems to modern haptic technologies, the evolution of tactile experiences highlights the dynamic nature of the field.

The latest technological advances with generative AI provide yet another tool for artists considering tactile elements. The integration of AI, such as OpenAI's ChatGPT 4 with vision capabilities, revolutionizes how neurodivergent individuals engage with tactile activities like crafting, offering new opportunities for sensory exploration and understanding. By taking a photograph of different knitting yarns with a smartphone and querying an AI like ChatGPT Vision about the sensory attributes of each yarn, neurodivergent individuals can make informed decisions about which materials best suit their sensory needs (Fig. 2.17). The AI can analyze textures, colors, and patterns, providing insights into the comfort and sensory impact of each type of yarn (Fig. 2.18). This not only personalizes the crafting experience but also empowers individuals to create in a way that is most comfortable and satisfying for them, enhancing their creative expression and sensory well-being. The potential of AI to interpret and advise on sensory-friendly materials represents a significant advancement in personalized crafting and neurodivergent accommodation, making creative pursuits more accessible and enjoyable. These advancements not only cater to specific sensory needs but also broaden our understanding of sensory perception, making the arts more inclusive and accessible.

Contemporary art is experiencing a notable fusion of tactile components to craft a multisensory experience that goes beyond conventional visual and auditory limitations. These artistic endeavors highlight the innovative ways in which tactile experiences and technologies are being utilized in the art world, expanding the traditional boundaries of sensory perception and artistic expression. The work of Bernhard Riecke stands as an example of integrating tactile sensory experiences with the realm of art, pushing the boundaries of how we engage with virtual and real environments (http://ispace.iat.sfu.ca/person/riecke/). As a psycho-physicist and cognitive scientist, Riecke's extensive background, with a PhD in Physics from Tübingen University and significant research in the Virtual Reality group at the Max Planck Institute for Biological Cybernetics, informs his pioneering approach. Now a professor at Simon Fraser University's School of Interactive Arts & Technology, Riecke's research transcends traditional disciplines, combining immersive virtual environments with multidisciplinary research approaches to explore human spatial cognition, orientation, and behavior. His work on enabling robust and

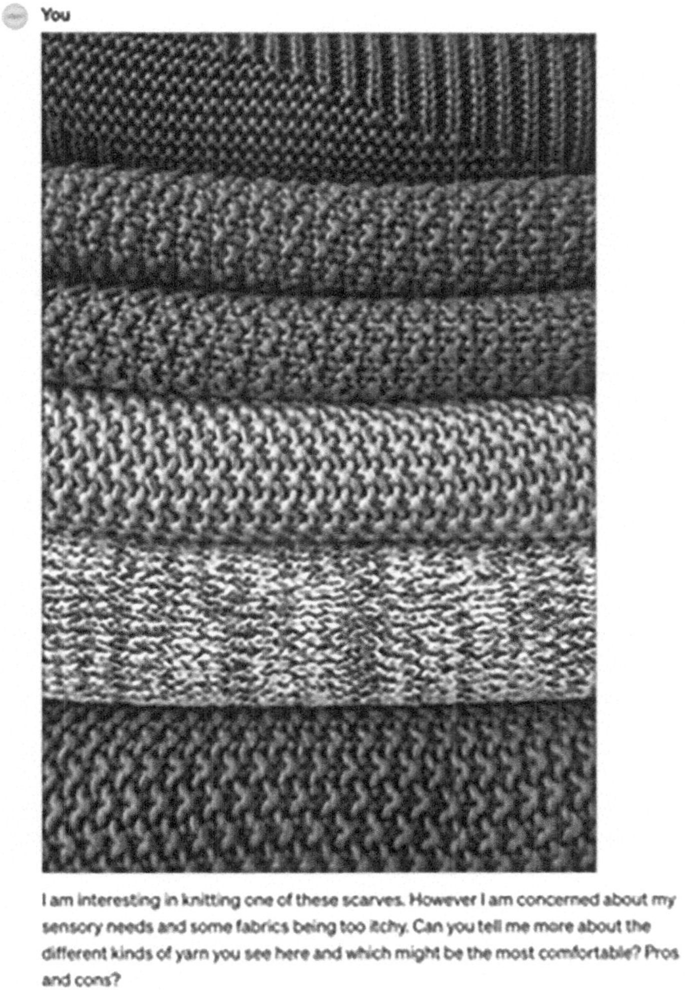

Fig. 2.17 Photograph of types of fabric with prompt. ChatGPT Vision, 2023

effortless spatial orientation in virtual environments, as well as his investigation into self-motion perception and multimodal cue integration, not only advances fundamental scientific knowledge but also guides the design of novel human–computer interfaces. These interfaces facilitate intuitive, embodied interactions within computer-mediated environments

ChatGPT

The image you provided shows a stack of knit scarves with varying textures and colors. While I can't provide tactile feedback or identify the specific yarn types from a photo, I can give you general advice based on the appearance of these knits.

From top to bottom:

1. **Teal Scarf (Topmost)**: This appears to be knit with a smooth, possibly synthetic or cotton-blend yarn. It has a tight knit, which suggests it may not be very stretchy but could be less itchy than wool. Synthetics and cotton are often good choices for sensitive skin.
   - *Pros*: Likely to be soft, less itchy, machine washable.
   - *Cons*: May not be as warm as wool, could pill over time.
2. **Green Scarf**: This one has a chunkier, textured knit, which looks like it could be wool or a wool blend. Wool is warm but can be itchy for sensitive skin, though merino wool or cashmere can be softer options.
   - *Pros*: Warm, good insulation, durable.
   - *Cons*: Potential itchiness, may require hand washing.
3. **Grey and White Scarf**: The yarn here has two colors twisted together, known as a marled effect. It's a tight knit, and the yarn looks smooth, possibly a cotton or acrylic blend.
   - *Pros*: Less likely to be itchy, durable, possibly machine washable.
   - *Cons*: May not be as warm as the green scarf, could feel heavy.
4. **Teal Scarf (Bottom)**: Similar to the top scarf, it's a smoother, tightly knit fabric that looks synthetic or cotton-blend.
   - *Pros*: Likely non-itchy, easy to care for, smooth against the skin.
   - *Cons*: May not breathe as well as natural fibers, could be less warm.

For sensitive skin, you might want to avoid rough wools and opt for softer, hypoallergenic yarns like bamboo, cotton, fine merino, silk, or acrylic. Always feel the yarn before buying if possible, or choose yarns labeled as 'soft' or 'skin-friendly'. If you're buying online, look for reviews that might give insight into the yarn's comfort. Remember, the tighter the knit, the less stretch and possibly the less breathability in the fabric, which can also affect comfort.

Fig. 2.18 ChatGPT response

like VR, enriching both the creation and experience of multimodal, interactive art and dance performances. Riecke's commitment to enhancing human–computer interaction reflects a deep understanding of the interconnectedness of sensory experiences, offering new vistas for artistic exploration and audience engagement.

Another example in this realm can be found in Alexandro Perini's *Tactile Headset* (2014) (https://alessandroperini.com/2014/09/11/tactileheadset/), a device that transmits information through vibrations. This headset provides an alternative form of sensory input for those with auditory or visual impairments, broadening the scope of sensory engagement in art. Similar environmental experiences use sound while reducing other stimuli. In her first major work titled *The Whispering Room* (1991), Canadian artist Janet Cardiff (1957–) introduced a captivating auditory experience in a minimalist artwork set in a dimly lit space, featuring sixteen compact, circular speakers mounted on stands. Each speaker emitted the voice of a distinct character, creating an immersive soundscape. As visitors navigated through this array of voices, their movements activated a film projector. This projector then displayed a film, subtly altered to play in slow motion, adding a visual layer to the auditory immersion. This work exemplified Cardiff's innovative approach to blending sound and visual elements, creating a multisensory environment that engaged visitors in a unique and interactive way.

Still, other artists akin to "cyborg artist" Neil Harbisson have integrated haptics directly into their bodies. Moon Ribas (1985–), for instance, a Spanish cyborg dancer and activist utilizes this technology to sense seismic activity in real time. Developing and implanting online seismic sensors in her feet allows her to experience earthquakes through vibrations from the ground. Thus, her work is a testament to the innovative ways artists are merging technology with human sensory experiences, offering new perspectives on how we interact with our environment (Alcaraz, 2019). Another artist also uses technology to enhance their work and bridges the gulf between olfactory and gustatory sensory studies. Blind artist Emile Gossiaux (1989–) (http://www.emiliegossiaux.com/), who, despite having no light perception, harnesses the unique capabilities of the BrainPort device (https://youtu.be/8Ccjq5LaSyE). The dense sensory nerve network of the tongue receives these electrical signals, which the brain interprets as visual images. A camera, mounted between the eyes, captures visual stimuli from the environment. This visual data is then processed into electrical pulses by an electronic unit. These pulses are conveyed to a plate of electrodes placed on the tongue. The dense sensory nerve network of the tongue receives these electrical signals, which the brain interprets as visual images. Thus, the BrainPort device allows individuals with visual impairments to perceive their surroundings in a novel way. As in the case of Gossiaux, she uses the device coupled with a rubber drawing board that generates tactile lines. The approach opens a new path into the olfactory and gustatory realms, where sensory experiences can be enhanced or even reimagined through technological intervention.

2.5 Olfactory and Gustatory Sensory Case Studies

The exploration of olfactory and gustatory sensory experiences delves into the intricate physiology and neurology of taste and smell. While considered biologically necessary, the significance of the senses goes beyond mere nourishment or warning of danger; they also elicit profound emotional connections, underpinned by complex neurological processes. These senses are deeply interconnected and play a vital role in shaping our perception and interaction with the world. The olfactory system, responsible for our sense of smell, is notably the most direct pathway to memory, owing to its close synaptic connection to the brain. Olfactory receptors in the nasal cavity capture scent molecules, which are then relayed to the olfactory cortex (Fig. 2.19). The process profoundly influences our emotions and memories, as scents are four times more likely to be remembered accurately than images and for longer durations (Olofsson et al., 2020). As such, smells have the ability to transport us past events or immediately call to mind the smell of individuals, leading to highly nostalgic experiences (Green et al., 2023).

The human olfactory system, with its capacity to detect approximately 1 trillion odors, is facilitated by an extensive array of over 400 types of scent receptors. These receptors, intriguingly, undergo renewal every 30 to 60 days, demonstrating the dynamic nature of our sense of smell. The olfactory receptors, located in the nasal cavity, play a crucial role in capturing scent molecules and transmitting them to the olfactory cortex in the temporal lobe of the brain (Gaillard et al., 2004). This process significantly influences our emotions and memory. The olfactory cortex's connection to regions like the amygdala and hippocampus is particularly noteworthy, as these areas are integral to emotional processing and memory formation. The activation of these regions by certain smells can have a profound impact, potentially reducing stress and lowering cortisol

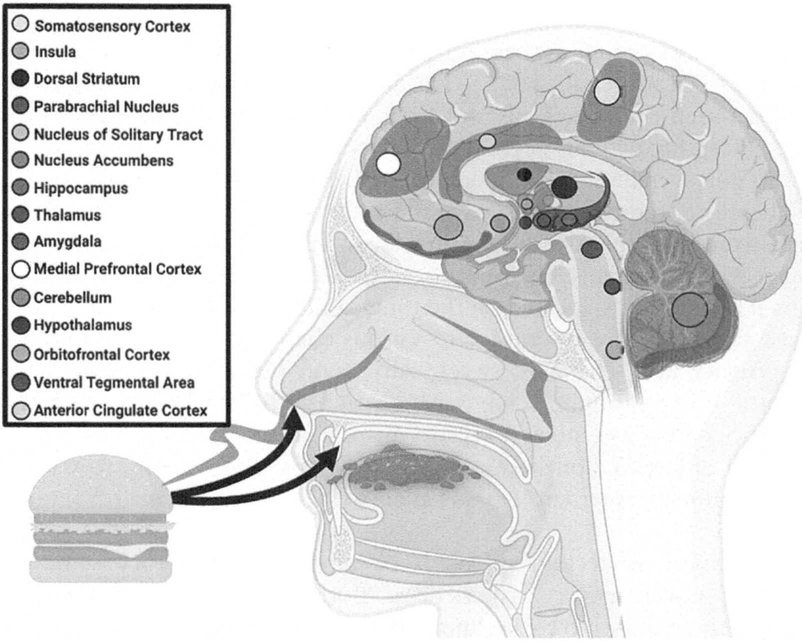

Fig. 2.19 Taste areas in the brain (*Source* From Wikimedia Commons, licensed under CC-BY 4.0)

levels, thus underscoring the therapeutic potential of olfactory stimulation (Duroux et al., 2020).

The significance of the olfactory bulb in this process cannot be overstated, as it is a unique component of the brain which continues to grow throughout our lives, generating new neurons and feeding directly into the limbic system (Shah et al., 2023). The limbic system encompasses the amygdala, hippocampus, and thalamus, and is pivotal in modulating emotions and managing the creation and consolidation of long-term memories (Thomasson et al., 2023). In parallel, the gustatory system, responsible for our sense of taste, is equally complex and vital. The human tongue houses approximately 10,000 taste buds, each containing receptors that respond to the five basic taste sensations: sweet, salty, sour, bitter, and umami. These taste buds are directly connected to the gustatory cortex in the brain, where the sensory information is processed and interpreted. The gustatory cortex not only deciphers the flavors we experience but also integrates these sensory inputs with emotional and visceral responses (Vicario et al., 2022). This intricate network of signaling pathways between the taste buds and the gustatory cortex underscores how taste perception extends beyond mere flavor recognition, playing a pivotal role in our overall sensory experience and emotional well-being.

Therefore, the olfactory system not only detects and processes scents, but also imbues them with emotional associations before they reach the upper cortex. Similarly, the sense of taste is governed by clusters of cells known as taste buds, located on the tongue and in the back of the mouth. These taste buds are responsive to various chemicals in food, discerning the primary taste channels: sweet, salty, sour, bitter, and umami. Each receptor on the tongue sends separate signals to the brain, where the gustatory cortex assembles a sensory map. This mapping enables experience and differentiation between various flavors, contributing to visceral and emotional experiences associated with eating (Schamarek et al., 2023).

The fusion of technology with olfactory and gustatory senses is unveiling novel sensory experiences. Both artists and scientists are utilizing these senses to create immersive and poignant experiences. These efforts stimulate taste and smell while also activating emotional and memory-related faculties, offering comprehensive sensory explorations. Technological advancements in the realm of olfaction have particularly noteworthy applications. For instance, the recognition of common smells is now a vital diagnostic tool in identifying neurodegenerative diseases. Individuals with memory loss, such as those suffering from Alzheimer's, often exhibit a significantly reduced sensitivity to smell. Addressing this, the Ode dementia player serves as a personal scent device for dementia patients, diffusing specific smells in a room to stimulate appetite, particularly around mealtimes (Yamashita et al., 2023). Additionally, Cyrano is a notable innovation in this field—a portable digital scent speaker that combats olfactory fatigue. This phenomenon refers to the gradual loss of awareness of a smell over time, which can lessen the sensory burden of the brain. With Cyrano, users can insert scent disks into the device and control the output through a smartphone app named oNotes (Kitson & McHugh, 2019). In 2007, the Brooklyn-based company Flavor Paper introduced the first scratch-and-sniff wall covering, marking a novel intersection between everyday objects and sensory stimulation. Such technological interventions demonstrate the growing importance and potential of integrating the senses of smell and taste with technology, paving the way for new and profound sensory explorations.

Such explorations include research into the effects of other senses on smell and vice versa. For instance, the intersections between different senses, particularly taste and smell, reveal the complex yet harmonious synergy of our sensory experiences. A notable study by Ward et al. (2023) at Liverpool John Moores University delved into how our sense of smell can subtly alter color perception. The research team discovered that specific odors, such as caramel and coffee, are subconsciously linked to certain colors, thereby influencing our perception of color. Participants were exposed to various scents in a sensory-deprived room and then asked to adjust a randomly colored square to a neutral gray. Intriguingly, while most scents led to predictable color perception shifts, peppermint was an exception, indicating a nuanced interplay of sensory associations. The study sheds light on how sensory inputs are not isolated but intricately woven into our daily perceptions and emphasizes the profundity of the brain's ability to process information from multiple senses simultaneously. For instance, higher temperatures are often associated with warmer colors, lower sound pitches with lower physical positions, and specific colors with the taste of certain foods, like oranges with the color orange.

Extending this exploration into the realm of music and taste, Mesz et al. (2011) conducted an experiment linking musical qualities with taste. Musicians were asked to improvise compositions in response to basic tastes like bitter, sweet, and salty. The analysis revealed that music created in response to a bitter taste often had slower, lower, and more softly legato notes. Similar correlations were observed for sweet and salty tastes, suggesting a profound connection between auditory and gustatory experiences. Together these studies illuminate the multifaceted and interconnected nature of our sensory world, highlighting how our perceptions are a blend of various sensory inputs. The insights gained from such research are crucial for advancing our understanding of sensory integration and its implications in fields like neuroarts and sensory therapy. As we continue to explore these sensory intersections, we uncover the vast potential for creating immersive and transformative experiences that engage multiple senses in harmony.

The realm of sensory art, particularly focusing on olfaction and gustation, has witnessed remarkable innovations, where artists have creatively integrated these senses into their works. These artistic endeavors not only offer unique sensory experiences but also serve as explorations into the emotional and cognitive impacts of smell and taste. One pioneering project in this domain was a collaboration between International Flavors and Fragrances, the Dutch design collective Polymorph, and Professor Asifa Majid. They conceived an "invisible dictionary" of contemporary emotional states, rooted in research on indigenous communities like the Jahai, who possess an elaborate smell vocabulary. The project, blending emotional states like collective *déjà vu* with original scents crafted by senior perfumer Le Guernec, was presented through an installation of translucent pedestals, releasing scents via laser-cut text. This immersive experience enabled visitors to forge new emotional-smell associations (Lupton & Lipps, 2018).

Another example of the use of smell in art comes by way of Sensory Maps from Kate McLean. These "smell maps" offer a captivating exploration of urban environments through the lens of olfactory experiences (https://sensorymaps.com/about/). Her innovative approach involves collecting and meticulously analyzing olfactory samples from diverse locations within cities. From these samples, McLean crafts maps that convey the unique narrative of each city, articulated not through traditional visual symbols but through its distinct smells. These "smell maps" represent an intersection of human-perceived smellscapes, cartography, and the communication of sensory data that typically eludes the eye.

In creating these sensory maps, McLean steps beyond conventional cartographic boundaries. She often omits traditional elements like street names, architectural landmarks, rivers, and parks, choosing instead to represent geographic space solely through the depiction of odors and their pathways. This method offers viewers an insightful and emotionally resonant exploration of cities, transcending the visual to engage with the often overlooked yet profoundly impactful world of urban scents. These maps challenge and redefine our understanding of spatial narratives, emphasizing the vital role of olfactory experiences in shaping our perception of and relationship with urban spaces.

The realm of "edible art" is epitomized by the innovative work of Sam Bompas, co-founder of Bompas & Parr (https://youtu.be/y03AI5Ad9IM). This studio, leading in flavor-based experience design, culinary research, architectural installations, and contemporary food design, consists of a team of creatives, cooks, designers, technicians, and architects. In one of their most remarkable installations, visitors find themselves immersed in an environment resembling a giant sandwich, complete with bread walls, lettuce ceilings, and tomato floors. But the explorations of the pair extend beyond mere culinary creations. For instance, their experiments with jelly demonstrate the transformative potential of scaling a simple concept to extraordinary levels. Their work reveals the unexplored possibilities within everyday items: a gherkin, ordinarily a mere food item, becomes a source of light when electricity is passed through it, emitting a sodium glow due to its vinegar content. In a striking installation from 2015, Bompas and Parr created a work at the intersection of taste and sight with *Gerkin Chandelier* (Fig. 2.20). The work used 60 pickles as light sources and consumed the electricity typical of an entire city block, creating a magnificent, albeit expensive, chandelier.

This edible installation is a multisensory feast, stimulating not just the senses of smell and taste, but also providing a visually and physically immersive experience. The work of Bompas & Parr and the other artists mentioned are emblematic of the diverse and innovative approaches artists are adopting to incorporate olfactory and gustatory elements in their art. These endeavors challenge and expand the traditional boundaries of artistic expression, offering fresh perspectives on sensory engagement and

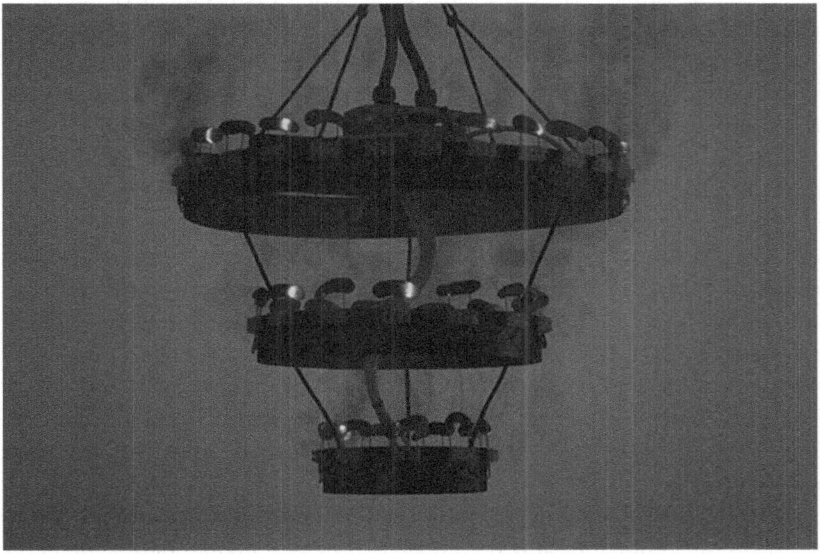

Fig. 2.20 Sam Bompas and Harry Parr, *Gherkin Chandelier*, 2015 (*Source* Permission of the artist)

interaction with the world. As we transition into discussing the future of sensory experiences in art, these examples stand as a testament to the limitless potential for creativity and exploration in the realm of neuroarts.

2.6 The Future of Sensory Experience in Art

As we envision the future of sensory experience in art, intersecting the realms of neurology, technology, and artistic creativity, the focus extends beyond merely capturing attention and conveying information. The emerging landscape emphasizes a profound engagement with our emotional and cognitive processes. This integration not only enhances the aesthetic experience but also fosters a deeper, more meaningful interaction with art, resonating with our psychological and emotional states. Thus, the future of sensory experience in art is intrinsically linked to our understanding of the neurochemical responses of the brain. Artistic experiences can trigger the release of various neurochemicals, hormones, and endorphins, eliciting powerful emotional reactions. These biochemical responses are not mere epiphenomena; they play a crucial role in how art affects us on a deep, emotional level. The enhancement of memory saliency through emotional responses to art is a key area of exploration. The heightened emotional engagement facilitated by art can lead to more vivid and lasting memories of the experience, underscoring the potential of art as a powerful tool for emotional and cognitive stimulation.

Another fascinating aspect of the future of sensory experience in art lies in the exploration of individual brain connectivity patterns and their influence on artistic preferences. Neuroscientific research is increasingly revealing our brains are wired in unique ways, leading to diverse responses to artistic stimuli. Further investigations will revolutionize the way we approach art, moving toward more personalized and tailored experiences resonating with individual neural patterns. One area of research likely to prove fruitful is the role played by the default mode network (Fig. 2.21), a key brain network associated with the neurological basis of the self. This network, active during periods of internal focus, daydreaming, and meaning-making, is integral to our experience of art (Menon, 2023). By engaging the default mode network, art can become a medium for introspection, self-reflection, and the construction of personal narratives. This engagement not only enriches the artistic experience, it also contributes to our understanding of the self and our place in the world.

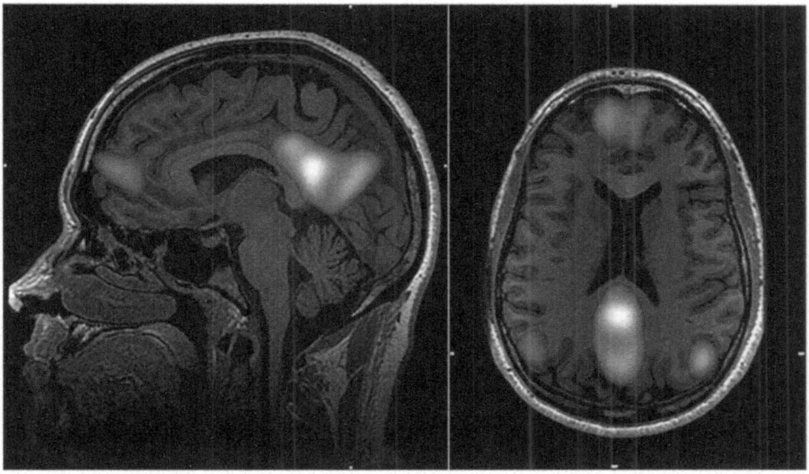

Fig. 2.21 Default mode network (*Source* From Wikimedia Commons, licensed under CC0)

Such research can be combined with new advances in technology to achieve the desired experiential states. The advent of sensory substitution devices offers a groundbreaking avenue for the future of sensory experience in art. Technologies crafted for the restoration or modification of specific sensory functions, or for converting one sensory experience into another, inaugurate avenues for enhancing accessibility and investigating sensory experiences (Eagleman & Perrotta, 2023). By harnessing these technologies, artists can create experiences which are inclusive and adaptable to diverse sensory capabilities, thus broadening the scope of who can engage with and appreciate art. Looking toward the future, we envision a landscape where artists leverage advanced technologies to create bespoke sensory experiences. Such experiences are poised not merely to stimulate the senses, but also to accommodate personal predilections and neural configurations. Among the foremost technologies harboring potential for these sought-after outcomes are those involving elements of sensory substitution.

In the realm of sensory substitution devices, innovations aimed at enhancing or replacing vision stand at the forefront, offering alternative perceptions to those with visual impairments. Of these, tactile cameras have garnered attention for their ability to translate visual information into tactile formats, thus enabling visually impaired individuals to interact more effectively with their environment (Maćkowski et al., 2023). One such device is the OrCam MyEye Pro, a wearable assistive technology providing smart reading, face recognition, color and product identification, and orientation assistance. It significantly aids individuals who are blind or visually impaired in navigating the visual world. Another notable innovation is the 2C3D tactile camera, created by Oren Geva, which converts images into 3D tactile representations, allowing blind individuals to "feel" images and gain a deeper understanding of the visual world (Matchinski et al., 2023).

Additionally, the development of "acoustic touch" technology in smart glasses marks a significant leap forward. This technology, pioneered by Australian researchers, translates visual data into unique auditory cues, significantly improving the ability of users to detect and reach objects. The acoustic touch technology in these smart glasses uses distinct sound representations to convey visual information, such as rustling for plants or buzzing for mobile phones, enhancing the independence and quality of life for the visually impaired (Zhu et al., 2023). Also, advancements in addressing colorblindness showcase the potential of sensory substitution devices in art experiences. Applications that allow real-time color correction offer individuals with colorblindness a more authentic and meaningful interaction with art. The concept of "smart art," which adapts its colors based on the viewer's color perception, represents another innovative approach, enabling art to dynamically respond to individual viewers' sensory needs (Recupero et al., 2023). These technological breakthroughs in sensory substitution devices not only enhance visual experiences for those with impairments but also open up new avenues for artistic exploration and appreciation, paving the way for a future where art is accessible and resonant for all.

Along with those for vision, the landscape of wearable technology is rapidly evolving, offering innovative solutions for mood regulation and emotional well-being (Devi et al., 2023). Among these pioneering devices is the Mood Sweater, a groundbreaking tool that gauges one's emotional state and physiological responses. The device utilizes the data it collects to offer personalized support, such as haptic feedback to mitigate anxiety or actionable suggestions like engaging in physical activity or social interaction. Its ability to tailor interventions makes it a powerful ally in managing mental health. Another noteworthy project in this domain is the Friendship Bench, which focuses on leveraging technology to provide support for individuals suffering from depression. Other devices for health and wellness include HeartMath's Inner Balance and emWave are pioneering new paths in heart-rate variability and stress management. These tools utilize biofeedback to synchronize heart rhythm patterns with emotional states, thereby facilitating stress reduction and emotional regulation. Their ability to align physiological and emotional well-being has made them indispensable in the quest for mental balance. The effectiveness of these devices in enhancing heart-rate variability exemplifies the significant impact biofeedback technology can have on personal health management (Lin et al., 2023).

Muse Headbands represent a critical development in this technological landscape, concentrating on brainwave activity to foster meditation and mindfulness practices. These headbands provide invaluable real-time feedback about the user's mental state, promoting relaxation and focus during meditation sessions. By offering insights into brain function, Muse Headbands have become essential for individuals aiming to deepen their mindfulness practices and achieve greater mental serenity and clarity (Cheng & Lin, 2023). Likewise, technology by Thync takes a unique approach to stress reduction and sleep enhancement, employing low-level electrical stimulation to directly influence neural activity. By targeting specific cranial nerves and brain regions, this wearable technology demonstrates the capacity to modulate mood and promote relaxation, showcasing the advanced capabilities of wearable technology in altering and improving mental states. Additionally, addressing mental health concerns such as depression, anxiety, and insomnia, the Fisher Wallace Stimulator utilizes cranial electrotherapy stimulation (CES) to deliver a gentle electrical current to the brain. This innovative approach helps regulate neurotransmitter levels, providing a novel method for alleviating symptoms associated with these mental health issues.

In the realm of holistic health monitoring, Spire Health Tags offers a comprehensive approach. These discreet wearable devices monitor key indicators like respiratory patterns and heart rate, providing users with insightful data on stress levels, sleep quality, and physical activity. They represent a holistic approach to health monitoring, delivering essential information that can guide lifestyle choices and enhance overall well-being. TouchPoints utilize bilateral alternating stimulation tactile (BLAST) technology, offering a novel method for managing stress and anxiety. By influencing the body's fight or flight response, these devices demonstrate the transformative potential of wearable technology in the realm of stress management and focus enhancement. Finally, Apollo Neuro employs gentle vibrations to improve the body's resilience to stress, illustrating the innovative use of haptic technology. This device impacts both the sympathetic and parasympathetic nervous systems, balancing the stress response and heralding a new age in wearable stress management technology. Each of these devices showcases the profound potential of technological innovation to revolutionize personal health and wellness, paving the way for a future where mental and emotional balance is within reach for everyone.

Diverse wearable devices represent a significant advancement in the intersection of technology, health, and well-being. By providing personalized interventions, real-time feedback, and innovative approaches to mental health management, they open up new possibilities for enhancing emotional regulation, reducing stress, and improving overall quality of life. As these technologies continue to evolve, they promise to redefine the way we approach personal health and emotional well-being.

At the same time, the intersection of art and technology, especially within the realm of neuroscience, heralds a new frontier in artistic expression. Including the pioneering initiative is Neurogami (https://neurogami.com/), which melds art with neuroscience. In this unique collaboration, artists and neuroscientists join forces to interpret fMRI brain scans, crafting artworks that echo the brain's activity. These creations range from direct visual representations of neural patterns to abstract works that embody the emotions or thoughts linked with the observed brain activity. This project exemplifies how the intricacies of the human brain can be translated into compelling artistic narratives.

Brain Painting (http://www.brainpaint.com/index.html) is another groundbreaking venture that intertwines art with brain science. In this project, participants don EEG headsets that capture brain activity as they engage with art. The EEG data then serves as the basis for generating new artworks, each uniquely reflective of the viewer's emotional and cognitive responses. This innovative approach personalizes art creation, making each piece a distinct representation of the individual's psychological and emotional state.

The BrainPort Vision Pro (https://www.wicab.com/brainport-vision-pro) represents a fascinating blend of art, neuroscience, and technology, primarily aimed at assisting individuals with vision loss. This device, worn on the tongue, translates visual information into electrical signals that the brain interprets as tactile sensations on the tongue. Over time, the brain learns to decode these signals as visual data, effectively teaching the brain to "see" through the tongue. This venture is a testament to the increasingly blurred boundaries between art, neuroscience, and technological innovation. Other devices, such as the Tactile-Vision Substitution System and Seeing with Sound, are transforming images into sensory experiences like vibrations on the skin or sound patterns. Though still experimental, these technologies show immense promise in redefining sensory perception and art creation.

Perhaps most importantly, Brain–Computer Interfaces (BCI) mark a significant leap in this interdisciplinary field. BCIs enable individuals to control computers through brain activity, opening a plethora of possibilities for those with physical disabilities. Artists are harnessing this technology to create unique forms of art. A notable example is musician Tod Machover (2004) (1953–), who collaborates with disabled individuals to compose music derived from their brain activity. This approach not only democratizes art creation but also illustrates the profound impact of integrating neuroscience and technology in the arts. Coupled with biofeedback, BCI promises new tailored experiences. At the forefront of this bespoke approach to experience is the integration of biofeedback into immersive art experiences. Researchers are exploring innovative ways to create art that dynamically responds to individual mental and physiological states (Nigg, 2023; Stern, 2010). In groundbreaking studies, participants engage with art while wearing devices that measure brain activity, allowing the art to morph in real time based on their emotional and cognitive responses (Prpa & Pasquier, 2019; Yang et al., 2023). Such an approach has proven potential to craft highly personalized art experiences which cater to individual preferences and also promote mental well-being.

Emerging technologies like haptic feedback are being developed to aid individuals with sensory processing disorders. This technology can gently acclimate individuals to challenging sensory inputs, such as loud noises or bright lights, in a controlled and therapeutic manner (Davies & Gavin, 2007). Such interventions hold immense promise for enhancing sensory tolerance and comfort in daily life for individuals with conditions like autism. The future of sensory experience in art is rich with possibilities, marked by a shift toward more inclusive, adaptive, and personally meaningful forms of artistic expression. However, the value of modulating sensory experiences hinges critically on their alignment with the neurotype of the intended audience. Understanding the diagnostic criteria and symptoms of each sensory processing condition is imperative to tailor the stimuli precisely, ensuring the outcome aligns with the artist's vision and the audience's sensory profile. This refined approach recognizes the variety in sensory processing among individuals and initiates contemplation on how the art experience can expand to become more inclusive and accessible.

The future of sensory experience in art, thus, is a balance between artistic intent and audience receptivity. It is a domain where art and technology converge with an understanding of neurological diversity, fostering experiences that are as unique and dynamic as the individuals interacting with them. This era of art challenges and redefines the traditional norms, paving the way for bespoke experiences able to honor and cater to the varied sensory and cognitive landscapes of its audience. The promise of this future lies in its ability to adapt art to individual needs, making it a deeply personal and engaging journey for each participant.

References

Abraira, V. E., & Ginty, D. D. (2013). The sensory neurons of touch. *Neuron, 79*(4), 618–639.

Akın, A. T., & Cömert, Ç. (2023). The development of an augmented reality audio application for visually impaired persons. *Multimedia Tools and Applications, 82*(11), 17493–17512.

Alcaraz, A. Ł. (2019). Cyborgs' perception, cognition, society, environment, and ethics: Interview with Neil Harbisson and Moon Ribas, 14 October 2016, Ace Hotel, New York City. *Journal of Posthuman Studies, 3*(1), 60–73.

Almenas, J. (2022, April 5). A canvas of umbrellas: Florida Women's Art Association opens spring exhibit at MOAS. *Observer.* https://www.observerlocalnews.com/news/2022/apr/05/a-canvas-of-umbrellas-florida-womens-art-association-opens-spring-exhibit-at-moas/

Andrieu, B., Bernard, A., Cipoletta, G., Marussich, Y., & da Nobrega, P. (2020). Emersive microperformativity: On physiological mediation in Yann Marussich's 'immobile'performances. *Performance Research, 25*(3), 137–144.

Ather, S., Proudlock, F. A., Welton, T., Morgan, P. S., Sheth, V., Gottlob, I., & Dineen, R. A. (2019). Aberrant visual pathway development in albinism: From retina to cortex. *Human Brain Mapping, 40*(3), 777–788.

Barrera, H. (2021, April 16). EEG Brainwave art: Giving a different meaning to our memories. *Medium.* https://medium.com/neurotech-davis/eeg-brainwave-art-giving-a-different-meaning-to-our-memories-202267dd2d55

Barrett, G. D. (2023). Technological catastrophe and the robots of Nam June Paik. *Cultural Critique, 118*(1), 56–82.

Bembibre, C., & Strlič, M. (2022). From smelly buildings to the scented past: An overview of olfactory heritage. *Frontiers in Psychology*, *12*, 718287.

Bergamasco, M., Frisoli, A., & Barbagli, F. (2002, June). Haptics technologies and cultural heritage applications. In *Proceedings Of Computer Animation 2002* (CA 2002) (pp. 25–32). IEEE.

Bergantini, L. P. (2019, October). Synesthesia: Meeting point between science, art and technology in the context of digital culture. In *Proceedings of the 9th International Conference on Digital and Interactive Arts* (pp. 1–7).

Bliss, T. K., Iwasaki, T., & Bart-Smith, H. (2012). Resonance entrainment of tensegrity structures via CPG control. *Automatica, 48*(11), 2791–2800.

Blundell, E. W. (2023). Multisensory visitor engagement. In *Museums and technologies of presence* (pp. 93–111). Routledge.

Bode, M. (2022). "Can you hear me now?": Listening and living in disruptive (post) pandemic soundworlds. In *Art-based research in the context of a global pandemic* (pp. 148–170). Routledge.

Boden, M. A., & Edmonds, E. A. (2019). *From fingers to digits: An artificial aesthetic*. MIT Press.

Borghesani, V., & Piazza, M. (2017). The neuro-cognitive representations of symbols: The case of concrete words. *Neuropsychologia, 105*, 4–17.

Bouvet, L., Amsellem, F., Maruani, A., Dupont, A. T. V., Mathieu, A., Bourgeron, T., Delorme, R., & Mottron, L. (2019). Synesthesia & autistic features in a large family: Evidence for spatial imagery as a common factor. *Behavioural Brain Research, 362*, 266–272.

Brackett, D. (2022). Outlier/indweller: Writing as walking. *Ought: The Journal of Autistic Culture, 4*(1), 10.

Bunt, L., & Stige, B. (2014). *Music therapy: An art beyond words*. Routledge.

Calzavarini, F. (2023). Rethinking modality-specificity in the cognitive neuroscience of concrete word meaning: a position paper. *Language, Cognition and Neuroscience*, 1–23.

Cardullo, R. J. (2017). Gesamtkunstwerk, synesthesia, and the avant-garde: Wassily Kandinsky's the yellow sound as a work of art. *Hermeneia: Journal of Hermeneutics, Art Theory & Criticism*, 18.

Castelló, M., Burguez, C., Pisani, M., & Umpiérrez, M. (2023). Biofeedback-driven sound and image generation. *IBRO Neuroscience Reports, 15*, S955–S956.

Cephas, A. S., Sofield, S., & Millstein, A. (2022). Embracing technological possibilities in the telehealth delivery of interactive music therapy. *Nordic Journal of Music Therapy, 31*(3), 214–227.

Chaudhury, A. N., Man, T., Yuan, W., & Atkeson, C. G. (2022). Using collocated vision and tactile sensors for visual servoing and localization. *IEEE Robotics and Automation Letters, 7*(2), 3427–3434.

Cheng, C. F., & Lin, C. J. (2023). Building a low-cost wireless biofeedback solution: Applying design science research methodology. *Sensors, 23*(6), 2920.

Chepaitis, A., Griffiths, F., Wyatt, H., & O'Connell, W. (2004). Evaluation of tactile fonts for use by a visually impaired elderly population. *Visual Impairment Research, 6*(2–3), 111–134.

Clark, A. (2006). Material symbols. *Philosophical Psychology, 19*(3), 291–307.

Cocozza, S. N. (2022). *Listening otherwise* (Doctoral dissertation, Tufts University).

Cox, S., & Guillemin, M. (2018). Enhancing meaning-making in research through sensory engagement with material objects. *The Qualitative Report, 23*(11), 2754–2771.

Cressman, J., & DeTora, L. (2021). Graphic embodiments: Perspectives on health and embodiment in graphic narratives. *Graphic Embodiments,* 1–210.

Dalton, B., & Musetti, K. (2018). Tactile picture book making and multimodal composition: Students design for equity in English language arts. In *Best practices in teaching digital literacies* (pp. 195–213). Emerald Publishing Limited.

Damayanti, T., Yuliana, Y. G. S., & Sudarsono, S. (2023). The dual-coding theory and digital media: The effect of nursery rhymes on teaching vocabulary. *Lectura: Jurnal Pendidikan, 14*(2), 165–174.

Davies, P. L., & Gavin, W. J. (2007). Validating the diagnosis of sensory processing disorders using EEG technology. *The American Journal of Occupational Therapy, 61*(2), 176–189.

Devi, D. H., Duraisamy, K., Armghan, A., Alsharari, M., Aliqab, K., Sorathiya, V., Das, S., & Rashid, N. (2023). 5g technology in healthcare and wearable devices: A review. *Sensors, 23*(5), 2519.

Dhar, S. K., & Das, M. (2020). Music in the time of COVID-19. *Biomedical Journal of Scientific & Technical Research, 31*(2), 24025–24028.

Daris, G. (2021). From an atmospheric point of view: Gustav Metzger in the context of the air war. *Performance Research, 26*(7), 154–162.

Droumeva, M. (2021). Soundscapes of productivity: The coffee-office and the sonic gentrification of work. *Resonance: The Journal of Sound and Culture, 2(*3), 377–394.

Duroux, R., Mandeau, A., Guiraudie-Capraz, G., Quesnel, Y., & Loing, E. (2020). A rose extract protects the skin against stress mediators: A potential role of olfactory receptors. *Molecules, 25*(20), 4743.

Eagleman, D. M., & Perrotta, M. V. (2023). The future of sensory substitution, addition, and expansion via haptic devices. *Frontiers in Human Neuroscience, 16*, 1055546.

Eardley, A. F., Mineiro, C., Neves, J., & Ride, P. (2016). Redefining access: Embracing multimodality, memorability and shared experience in Museums. *Curator: The Museum Journal, 59*(3), 263–286.

Engeset, R. V., Pfuhl, G., Orten, C., Hendrikx, J., & Hetland, A. (2022). Colours and maps for communicating natural hazards to users with and without colour vision deficiency. *International Journal of Disaster Risk Reduction, 76*, 103034.

Fallon, M. (2014). *Creating the future: Art and Los Angeles in the 1970s*. Catapult.

Fenlon, I. (2002). *Music and culture in late renaissance Italy*. Oxford University Press.

Ferreri, L., Mas-Herrero, E., Zatorre, R. J., Ripollés, P., Gomez-Andres, A., Alicart, H., Olivé, G., Marco-Pallarés, J., Antonijoan, R. M., Valle, M., Riba, J., & Rodriguez-Fornells, A. (2019). Dopamine modulates the reward experiences elicited by music. *Proceedings of the National Academy of Sciences, 116*(9), 3793–3798.

Fraile, E., Gagnepain, P., Eustache, F., Groussard, M., & Platel, H. (2023). Musical experience prior to traumatic exposure as a resilience factor: A conceptual analysis. *Frontiers in Psychology, 14*, 1220489.

Fraisse, V., Schütz, N., Guastavino, C., Wanderley, M., & Misdariis, N. (2023, February). Informing sound art design in public space through soundscape simulation. In *INTER-NOISE and NOISE-CON Congress and Conference Proceedings* (Vol. 265, No. 4, pp. 3015–3024). Institute of Noise Control Engineering.

Frisbie, C. J. (1980). Vocables in Navajo ceremonial music. *Ethnomusicology, 24*(3), 347–392.

Fulas, T. D. A. (2023). The pioneer in the education of the blind, and deafblind in the United States: Samuel Gridley Howe (1801–1876). *Educar Em Revista, 39*, e87444.

Gaillard, I., Rouquier, S., & Giorgi, D. (2004). Olfactory receptors. *Cellular and Molecular Life Sciences CMLS, 61*, 456–469.

Ganis, G., & Kutas, M. (2003). An electrophysiological study of scene effects on object identification. *Cognitive Brain Research, 16*(2), 123–144.

Garofolo, I. (2022, September). Unidirectional tactile paving: Circulation for the visually impaired. In *Transforming our world through universal design for human development: Proceedings of the Sixth International Conference on Universal Design (UD2022)* (Vol. 297, p. 375). IOS Press.

Gibson, J. J. (2002). A theory of direct visual perception. *Vision and Mind: Selected Readings in the Philosophy of Perception*, 77–90.

Gold, B. P., Mas-Herrero, E., Zeighami, Y., Benovoy, M., Dagher, A., & Zatorre, R. J. (2019). Musical reward prediction errors engage the nucleus accumbens and motivate learning. *Proceedings of the National Academy of Sciences, 116*(8), 3310–3315.

Goller, A. I., Otten, L. J., & Ward, J. (2009). Seeing sounds and hearing colors: An event-related potential study of auditory–visual synesthesia. *Journal of Cognitive Neuroscience, 21*(10), 1869–1881.

Green, T., Hilkhuysen, G., Huckvale, M., Rosen, S., Brookes, M., Moore, A., Naylor, P., Lightburn, L., & Xue, W. (2022). Speech recognition with a hearing-aid processing scheme combining beamforming with mask-informed speech enhancement. *Trends in Hearing, 26.*23312165211068629

Green, J. D., Reid, C. A., Kneuer, M. A., & Hedgebeth, M. V. (2023). The proust effect: Scents, food, and nostalgia. *Current Opinion in Psychology, 50*, 101562.

Guga, J., & Uspenski, I. (2016). Procedural aesthetics and the emergence of NeuroArt. In *Going digital: Innovations in art, architecture, science and technology—Book of abstracts* (pp. 40–41).

Harbisson, N. (2019). Hearing colors: My life experience as a cyborg. In *Creativity, imagination and innovation: perspectives and inspirational stories* (pp. 117–125).

Harkness, S. S., Noblitt, B. A., & Giesbers, N. (2023). Synesthesia: 3.1415... Orange. WhitePeriwinkleWhiteBlue... *Journal of Humanistic Mathematics, 13*(2), 351–367.

Heddaya, M. (2013, November 1). The sounds of silence. *Hyperallergic.* https://hyperallergic.com/91011/the-sounds-of-silence/

Heller, G. N. (1987). Ideas, initiatives, and implementations: Music therapy in America, 1789–1848. *Journal of Music Therapy, 24*(1), 35–46.

Hennessy, S., Sachs, M., Kaplan, J., & Habibi, A. (2021). Music and mood regulation during the early stages of the COVID-19 pandemic. *PLoS ONE, 16*(10), e0258027.

Hervás, R., Bautista, S., Méndez, G., Galván, P., & Gervás, P. (2020). Predictive composition of pictogram messages for users with autism. *Journal of Ambient Intelligence and Humanized Computing, 11*(11), 5649–5664. https://doi.org/10.1007/s12652-020-01925-z

Hoffmann, M. B., & Dumoulin, S. O. (2015). Congenital visual pathway abnormalities: A window onto cortical stability and plasticity. *Trends in Neurosciences, 38*(1), 55–65.

Holmström, I. (2022). Communication, information, and support for Swedish parents with deaf or hard-of-hearing children. *Scandinavian Journal of Disability Research, 24*(1), 165–180.

Hsiang, E. L., Yang, Z., Yang, Q., Lai, P. C., Lin, C. L., & Wu, S. T. (2022). AR/VR light engines: Perspectives and challenges. *Advances in Optics and Photonics, 14*(4), 783–861.

Huang, Y., Yao, K., Li, J., Li, D., Jia, H., Liu, Y., Chun Ki Yiu, C., Park, W., & Yu, X. (2022). Recent advances in multi-mode haptic feedback technologies towards wearable interfaces. *Materials Today Physics, 22*, 100602.

Hutmacher, F. (2019). Why is there so much more research on vision than on any other sensory modality? *Frontiers in Psychology, 10*, 2246.

Hutson, J., & Olsen, T. (2022). Virtual reality and learning: A case study of experiential pedagogy in art history. *Journal of Intelligent Learning Systems and Applications*, 14.

Imasato, N., Miyazawa, K., Duncan, C., & Nagai, T. (2023). Using a language model to generate music in its symbolic domain while controlling its perceived emotion. *IEEE Access*.

Khan, F. S., Van de Weijer, J., & Vanrell, M. (2012). Modulating shape features by color attention for object recognition. *International Journal of Computer Vision, 98*, 49–64.

Kim, J. S., & Kim, C. H. (2014). A review of assistive listening device and digital wireless technology for hearing instruments. *Korean Journal of Audiology, 18*(3), 105.

Kitson, J., & McHugh, K. (2019). Olfactory attunements and technologies: Exposing the affective economy of scent. *GeoHumanities, 5*(2), 533–553.

Lane, R. (2023). Of sight, and insight. *The Lancet, 401*(10371), 97.

Lee, K. K. (2023, November). Affective generative visuals based on data input influenced by user's emotions. In *Proceedings of the International Conference of Innovation in Media and Visual Design (IMDES 2023)* (Vol. 790, p. 387). Springer Nature.

León-Sanz, P. (2016). Evolution of the concept of emotion in medicine: A music therapy approach 1. In *The emotions and cultural analysis* (pp. 69–95). Routledge.

Lim, T. S. (2019). Moving meditation: Paik Nam June's *TV Buddha* and its zen buddhist aesthetic meaning. *Dao, 18*, 91–107.

Lin, I. M., Chen, T. C., Tsai, H. Y., & Fan, S. Y. (2023). Four sessions of combining wearable devices and Heart Rate Variability (HRV) biofeedback are needed to increase HRV indices and decrease breathing rates. *Applied Psychophysiology and Biofeedback, 48*(1), 83–95.

Lord, S. (2008). *Music in the Middle Ages: A reference guide*. Bloomsbury Publishing USA.

Lozano-Hemmer, R. (2010, June 18). Rafael Lozano-Hemmer. *Radio Program. Broadcast*.

Lupton, E., & Lipps, A. (Eds.). (2018). *The senses: Design beyond vision*. Chronicle Books.

Machover, T. (2004). Shaping minds musically. *BT Technology Journal, 22*(4), 171–179.

Maćkowski, M., Brzoza, P., Kawulok, M., Meisel, R., & Spinczyk, D. (2023). Multimodal presentation of interactive audio-tactile graphics supporting the perception of visual information by blind people. *ACM Transactions on Multimedia Computing, Communications and Applications, 19*(5s), 1–22.

Magsamen, S., & Ross, I. (2023). *Your brain on art: How the arts transform us*. Random House.

Martínez-Santiago, F., García-Viedma, M. R., Williams, J. A., Slater, L. T., & Gkoutos, G. V. (2020). Aging neuro-behavior ontology. *Applied Ontology, 15*(2), 219–239.

Matchinski, T. L., Crumbliss, K. E., Corgiat, E., & Pang, Y. (2023). Near prescribing trends in two low vision rehabilitation clinics over a ten-year period. *Clinical and Experimental Optometry*, 1–8. https://doi.org/10.1080/08164622.2023.2246490

McGrath, B., & Brennan, M. A. (2020). Tradition, cultures, and communities: Exploring the potentials of music and the arts for community development in Appalachia. In *Culture, community, and development* (pp. 147–169). Routledge.

Menon, V. (2023). 20 years of the default mode network: A review and synthesis. *Neuron*.

Mesz, B., Trevisan, M. A., & Sigman, M. (2011). The taste of music. *Perception, 40*(2), 209–219.

Monti, J. (2012). *A sense of the sacred: Roman Catholic worship in the Middle Ages*. Ignatius Press.

Morash, V. S., Pensky, A. E. C., Tseng, S. T., & Miele, J. A. (2014). Effects of using multiple hands and fingers on haptic performance in individuals who are blind. *Perception, 43*(6), 569–588.

Moreno-Artero, E., Morice-Picard, F., Bremond-Gignac, D., Drumare-Bouvet, I., Duncombe-Poulet, C., Leclerc-Mercier, S., Dufresne, H., Kaplan, J., Jouanne, B., Arveiler, B., Taieb, A., & Hadj-Rabia, S. (2021). Management of albinism: French guidelines for diagnosis and care. *Journal of the European Academy of Dermatology and Venereology, 35*(7), 1449–1459.

Morrison, I. (2023). Touching to connect, explore, and explain: How the human brain makes social touch meaningful. *The Senses and Society, 18*(2), 92–109.

Müller, K., Engel, C., Loitsch, C., Stiefelhagen, R., & Weber, G. (2022). Traveling more independently: A study on the diverse needs and challenges of people with visual or mobility impairments in unfamiliar indoor environments. *ACM Transactions on Accessible Computing (TACCESS), 15*(2), 1–44.

Murao, M. (2022). *Designing immersive art experience-an exploration of visuals and sounds* (Doctoral dissertation, Massachusetts Institute of Technology).

Nigg, J. T. (2023). Considerations toward an epigenetic and common pathways theory of mental disorder. *Journal of Psychopathology and Clinical Science, 132*(3), 297.

Nikolsky, A. (2020). The pastoral origin of semiotically functional tonal organization of music. *Frontiers in Psychology, 11*, 1358.

Oduntan, A. O., Raliavhegwa, M., & Lund, P. M. (2002). Vision status of children with oculocutaneous albinism. *Ophthalmic and Physiological Optics, 22*(6), 572–573.

Olivas, A. M. (2023). Los neoliberarchivos de Teresa Margolles: Contemporary Art at the Mexico-US Border. *Latin American and Latinx Visual Culture, 5*(2), 57–71.

Olofsson, J. K., Ekström, I., Lindström, J., Syrjänen, E., Stigsdotter-Neely, A., Nyberg, L., Jonsson, S., & Larsson, M. (2020). Smell-based memory training: Evidence of olfactory learning and transfer to the visual domain. *Chemical Senses, 45*(7), 593–600.

Othman, B. A., Ong, J. E., & Dumitrescu, A. V. (2022). Biallelic Optic Atrophy 1 (OPA1) related disorder—Case report and literature review. *Genes, 13*(6), 1005.

Pant, U., Frishkopf, M., Park, T., Norris, C. M., & Papathanassoglou, E. (2022). A neurobiological framework for the therapeutic potential of music and sound interventions for post-traumatic stress symptoms in critical illness survivors. *International Journal of Environmental Research and Public Health, 19*(5), 3113.

Phillips, R. (2015). Playful and multi-sensory fieldwork: Seeing, hearing and touching New York. *Journal of Geography in Higher Education, 39*(4), 617–629.

Pomberger, B. M., Kotova, N. S., & Stadler, P. (2021). New Hypothesized musical instruments of the European neolithic. *Arheologia, 1*(3), 28–35.

Prendergast, L. (2021). Blood lines (2014): United in diversity: 'But who could resist you dad?' In *Conscious theatre practice* (pp. 95–110). Brill.

Provenza, A. (2020). Music and medicine. In *A companion to Ancient Greek and Roman Music* (pp. 351–363). Wiley.

Prpa, M., & Pasquier, P. (2019). *Brain-computer interfaces in contemporary art: A state of the art and taxonomy* (pp. 65–115). Brain-Computer Interfaces for Artistic Expression.

Puts, N. A., & Cascio, C. J. (2023). Atypical development of tactile processing. In *Somatosensory research methods* (pp. 227–250). Springer US.

Qi, H., Yi, B., Suresh, S., Lambeta, M., Ma, Y., Calandra, R., & Malik, J. (2023, December). General in-hand object rotation with vision and touch. In *Conference on Robot Learning* (pp. 2549–2564). PMLR.

Rauscher, F. H., & Hinton, S. C. (2006). The Mozart effect: Music listening is not music instruction. *Educational Psychologist, 41*(4), 233–238.

Recupero, A., Marti, P., Regal, G., & Sackl, A. (2023). *Service co-design to envision the transformation of museums.*

Riby, L. M., Fenwick, S. K., Kardzhieva, D., Allan, B., & McGann, D. (2023). Unlocking the beat: Dopamine and eye blink response to classical music. *NeuroSci, 4*(2), 152–163.

Sacharin, V., Schlegel, K., & Scherer, K. R. (2012). Geneva emotion wheel rating study. *Center for Person, Kommunikation, Aalborg University, NCCR Affective Sciences. Aalborg University, Aalborg.*

Saifman, J., Colverson, A., Prem, A., Chomiak, J., & Doré, S. (2023). Therapeutic potential of music-based interventions on the stress response and neuroinflammatory biomarkers in COVID-19: A review. *Music & Science, 6*, 20592043221135810.

Santhosh, A. K., Sangilirajan, M., Nizar, N., Radhamani, R., Kumar, D., Bodda, S., & Diwakar, S. (2020). Computational exploration of neural dynamics underlying music cues among trained and amateur subjects. *Procedia Computer Science, 171*, 1839–1847.

Schamarek, I., Anders, L., Chakaroun, R. M., Kovacs, P., & Rohde-Zimmermann, K. (2023). The role of the oral microbiome in obesity and metabolic disease: Potential systemic implications and effects on taste perception. *Nutrition Journal, 22*(1), 1–13.

Sexton, J. (2021). Basic black. *Liquid Blackness, 5*(2), 75–83.

Shah, A., Jhawar, S. S., & Goel, A. (2023). Anatomy of the limbic system and surgery for medial temporal lesions. In *Functional anatomy of the brain: A view from the surgeon's eye* (pp. 285–298). Springer Nature Singapore.

Shi, W. Y. (2020). Re-examining the Mozart effect: The Sonata in D Major, K. 448 and the influence of rhythm on spatial intelligence. *Journal of Multidisciplinary Research, 12*(2), 121–129.

Shukla, V. K., & Verma, A. (2019, March). Enhancing user navigation experience, object identification and surface depth detection for "Low Vision" with proposed electronic cane. In *2019 Advances in Science and Engineering Technology International Conferences (ASET)* (pp. 1–5). IEEE.

Smardzewski, J., Kamisiński, T., Dziurka, D., Mirski, R., Majewski, A., Flach, A., & Pilch, A. (2015). Sound absorption of wood-based materials. *Holzforschung, 69*(4), 431–439.

Stern, D. N. (2010). *Forms of vitality: Exploring dynamic experience in psychology, the arts, psychotherapy, and development*. Oxford University Press.

Tai, S. K., & Kuo, Y. J. (2019, October). Alterations in brainwaves caused by different Music Genres. In *2019 IEEE 10th International Conference on Awareness Science and Technology (iCAST)* (pp. 1–4). IEEE.

Tao, W. W., Jiang, H., Tao, X. M., Jiang, P., Sha, L. Y., & Sun, X. C. (2016). Effects of acupuncture, Tuina, Tai Chi, Qigong, and traditional Chinese medicine five-element music therapy on symptom management and quality of life for cancer patients: A meta-analysis. *Journal of Pain and Symptom Management, 51*(4), 728–747.

Teeter, E. (2011). *Religion and ritual in ancient Egypt*. Cambridge University Press.

Tharpe, A. M., Eiten, L., & Gabbard, S. A. (2008, May). Hearing technology. In *Seminars in Hearing* (Vol. 29, No. 02, pp. 169–177). © Thieme Medical Publishers.

Thomasson, M., Voruz, P., Cionca, A., Jacot de Alcântara, I., Nuber-Champier, A., Allali, G., Benzakour, L., Lalive, P. H., Lövblad, K.-O., Braillard, O., Nehme, M., Coen, M., Serratrice, J., Reny, J.-L., Pugin, J., Guessous, I., Landis, B. N., Griffa, A., Van De Ville, D., Assal, F., & Péron, J. A. (2023). Markers of limbic system damage following SARS-CoV-2 infection. *Brain Communications, 5*(4), fcad177.

Thompson, P., & Christian, C. J. (2022). Louis Braille. *The Palgrave handbook of educational thinkers* (pp. 1–12). Springer International Publishing.

Tian, Z. (2023, September). Post-merge carbon footprint analysis and sustainability in the NFT Art Market. In *Arts* (Vol. 12, No. 5, p. 211). MDPI.

Towers, P. (2019). The Sound of an imitated ocean. *Cream City Review, 43*(1), 12–21.

Tuthill, J. C., & Azim, E. (2018). Proprioception. *Current Biology, 28*(5), R194–R203.

Tsioulcas, A. (2020, November 10). Struck with memory loss, a dancer remembers 'Swan Lake.' But who is she? NPR: Culture. https://www.npr.org/2020/11/10/933387878/struck-with-memory-loss-a-dancer-remembers-swan-lake-but-who-is-she

Ueno, F., & Shimada, S. (2023). Inter-subject correlations of EEG reflect subjective arousal and acoustic features of music. *Frontiers in Human Neuroscience*, 17.

University of Waterloo. (2023, August 22). Why our brains prefer symbols to words. *Neuroscience*. https://neurosciencenews.com/world-symbol-preference-23808/#:~:text=The%20research%20conducted%20by%20the%20University%20of%20Waterloo,which%20can%20have%20multiple%20meanings%2C%20enhance%20their%20memorability

Van Leeuwen, T. M., Wilsson, L., Norrman, H. N., Dingemanse, M., Bölte, S., & Neufeld, J. (2021). Perceptual processing links autism and synesthesia: A co-twin control study. *Cortex, 145*, 236–249.

Vicario, C. M., Nitsche, M. A., & Avenanti, A. (2022). Tongue motor cortex: The back door of the reward system. *Neuroscience & Biobehavioral Reviews*, 104820.

Vital, R., Sylaiou, S., Koukopoulos, D., Koukoulis, K., Dafictis, P., & Fidas, C. (2023). Comparison of extended reality platforms and tools for viewing and exhibiting art. *Digital Applications in Archaeology and Cultural Heritage, 31*, e00298.

Ward, R. J., Ashraf, M., Wuerger, S., & Marshall, A. (2023). Odors modulate color appearance. *Frontiers in psychology, 14*, 1175703.

Yamashita, Y., Shoji, Y., Yanagimoto, H., Morita, K., Kodama, H., Tsuruhisa, Y., & Ookawa, J. (2023). A characteristic of olfactory function in four types of dementia and non-dementia subjects using smell identification test. *Psychogeriatrics*.

Yang, X., Cheng, P. Y., Liu, X., & Shih, S. P. (2023). The impact of immersive virtual reality on art education: A study of flow state, cognitive load, brain state, and motivation. *Education and Information Technologies*, 1–20.

Zhan, X., Chen, C., Niu, L., Du, X., Lei, Y., Dan, R., Wang, Z.-W., & Liu, P. (2023). Locomotion modulates olfactory learning through proprioception in C. elegans. *Nature Communications, 14*(1), 4534.

Zhang, C., & Tan, G. (2023). Quantifying architectural color Quality: A Machine learning integrated framework driven by quantitative color metrics. *Ecological Indicators, 157*, 111237.

Zhao, Y., Bennett, C. L., Benko, H., Cutrell, E., Holz, C., Morris, M. R., & Sinclair, M. (2018, April). Enabling people with visual impairments to navigate virtual reality with a haptic and auditory cane simulation. In *Proceedings of the 2018 CHI conference on human factors in computing systems* (pp. 1–14).

Zhu, H. Y., Hossain, S. N., Jin, C., Singh, A. K., Nguyen, M. T. D., Deverell, L., Nguyen, V., Gates, F. S., Gorordo Fernandez, I., Vergel Melencio, M., Renee Bell, J.-A., Lin, C.-T., & Lin, C. T. (2023). An investigation into the effectiveness of using acoustic touch to assist people who are blind. *Plos one, 18*(10), e0290431.

PART III

Toward Personalization

CHAPTER 3

Sensory Profiles and Technological Deficits: Classification, Relevance, and Gaps

The chapter provides an analysis of the diverse range of sensory experiences and the challenges in addressing them with current technology. Opening with a discussion on how cultural and spiritual beliefs shape sensory perceptions, the chapter uses insights from various cultures, including the Berinmo people. It then delves into various aesthetic disabilities beyond visual perception, such as face blindness, dyschromatopsia, dyscalculia, and synesthesia, exploring their unique impacts on artistic interpretation. The chapter also tackles cross-modal sensory challenges, highlighting conditions like anosmia and agnosia, and their implications for experiencing art. Emphasizing neurodiversity, it examines how neurodivergent individuals, particularly those on the spectrum, have distinct sensory preferences and responses to art. Aphantasia's role in artistic experience is discussed, exploring how the absence of mental imagery influences art perception and creation. Finally, the chapter identifies key gaps in current technological applications, discussing issues like ableism in sensory technologies and the need for more inclusive and adaptable tools. It concludes by exploring therapeutic and adaptive applications, including color therapy and sensory adaptations in museums, emphasizing the importance of developing technologies that cater to a broad spectrum of sensory experiences and preferences. This chapter provides

J. Hutson et al., *Sensemaking and Neuroaesthetics*,
https://doi.org/10.1007/978-3-031-58045-1_3

crucial insights into the intricate relationship between sensory experiences, neurodiversity, and the role of technology in enhancing artistic engagement.

3.1 Introduction: Diverse Sensory Profiles in Personalization and Technological Application

The exploration of sensory diversity within the realms of neuroscience and technological advancements in the arts reveals the true diversity of human sense experience. While previous studies on synesthesia assumed the unique nature of sensory processing in individuals with sensory crossovers, the most recent research has begun to reveal that the human brain is, in fact, wired with parallel pathways which leads to the crosspollination of neural pathways (Fig. 3.1). Neural pathways in the brain are like intricate networks through which information travels. Neurons, the basic units of the brain, communicate with each other through these pathways. Each neuron can connect to thousands of others, forming a vast and complex network. When a neuron is activated, it sends an electrical signal down its axon to the synapse, where it releases chemical messengers called neurotransmitters. These neurotransmitters cross the synapse and bind to receptors on the next neuron, continuing the signal (Gopinath, 2023).

The neural pathways of the brain are responsible for everything from reflexive actions to complex thoughts, enabling sensory perception, movement, emotions, and cognition (Gulnoza et al., 2023). These pathways can also strengthen or reorganize through a process called neuroplasticity, especially in response to learning or after injury, demonstrating the remarkable ability of the brain to adapt (Alashram & Annino, 2023). Traditionally, neural pathways in the brain were thought to operate in a predominantly unidirectional manner, with signals flowing in a linear path from one neuron to the next. The view suggested a relatively straightforward mechanism of information transmission, akin to a one-way street (Courte et al., 2023). However, recent advancements in neuroscience have challenged this perspective, revealing a far more complex and dynamic network.

Emerging research from the École Polytechnique Fédérale de Lausanne (EPFL) has shed light on this unique aspect of human brain communication networks, revealing the transmission of information via multiple parallel pathways—a trait not observed in other mammals such as macaques or mice (Griffa et al., 2023). The discovery, made through

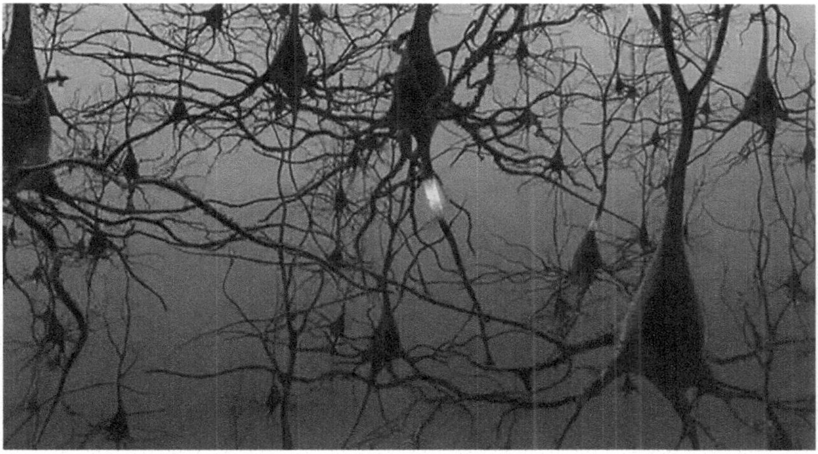

Fig. 3.1 Neural signaling (*Source* From Wikimedia Commons, licensed under CC0)

a fusion of diffusion and functional magnetic resonance imaging (MRI) data, along with information and graph theory, suggests these parallel pathways might be integral to our advanced cognitive abilities and could offer new insights into brain evolution and medical applications. This revelation underscores a paradigm shift in our understanding of brain functionality, highlighting the adaptable and interactive nature of neural pathways and suggesting a more integrated and interconnected neural landscape (Li, Xu et al., 2023).

The recent paradigm shifts in understanding neural pathways, from unidirectional conduits to complex, bidirectional networks, dovetails with the burgeoning field of creative neurotechnologies. The new understanding should be integrated into existing research which highlights the potential of such technologies to uniquely engage in neuroscience research, particularly in enhancing sensory experiences through artistic media (Aru et al., 2023; Calzavarini, 2023; Fivush & Grysman, 2023; Vázquez-Guardado et al., 2020). The interconnected nature of neural pathways underscores a more profound potential for art and technology to collaborate, illuminating the diverse sensory world with greater nuance and depth. In this evolving landscape, art therapy becomes increasingly significant, offering opportunities to rebalance brain functions compromised by various factors and enhance sensory perception and cognitive understanding (Kapitan, 2014). The progression in our understanding of brain functionality and sensory experience enriches the therapeutic potential of art, emphasizing the need for inclusive technologies that accommodate the full spectrum of neurodiversity (Stallings, 2022). The integration of these insights from neuroscience with the inventive applications of neurotechnology and art therapy represents a forward-looking approach, aiming to cater to and celebrate the diverse sensory experiences of individuals.

At the same time, the expansion of the technological frontier beyond therapy into realms like perceptual learning and sensory substitution devices represents a significant leap in altering and enhancing sensory experiences (Eagleman & Perrotta, 2023). The development of devices (discussed in Chapter 2) which allow for alternate ways of "seeing" or experiencing the world (Maimon et al., 2022; Proulx et al., 2014) underscores the immense potential of technological interventions. These innovations transcend the simple compensation for sensory disabilities; they revolutionize the essence of perception itself. Far from being mere replacements of one sense with another, they signify an expansion of the horizons of human perception and interaction with the environment.

As the interplay between neurological diversity and technological interventions is considered, it is crucial to ensure that these innovations are developed with inclusivity at their core. Inclusive sensory ethnography and similar approaches advocate for a design philosophy considering the full range of human sensory and cognitive experiences (Alper, 2018). This means creating technologies that do not just "add" to the sensory experience, but are adaptable and responsive to the unique ways individuals perceive and understand the world (Hamilton-Fletcher et al., 2022). By embracing both the complex nature of brain function and harnessing the transformative power of technology, we can envisage a future where each individual has the opportunity to perceive and engage with the world in the most personally meaningful ways. This vision requires both technological innovations and a commitment to understanding and valuing the diverse ways people experience the world, discovering new possibilities for connection, creativity, and understanding.

Before delving deeper into these specific conditions and technological interventions, it is imperative to understand the variability of sense experience by considering the cultural and ontological dimensions of sensory perception. These dimensions provide a crucial context for appreciating the full spectrum of sensory experiences and tailoring interventions that are culturally sensitive and personally meaningful.

They also significantly impact both the experience of and engagement with art and therapy, particularly when considering diverse sensory profiles (Rosler & Press-Barnathan, 2023). These dimensions of sensory perception refer to how sensory experiences and understandings are shaped by cultural context and the nature of being, respectively. Culturally, sensory perception is influenced by the values, beliefs, and practices of a particular community, leading to diverse interpretations and engagements with the sensory world (Arshamian et al., 2022). Ontologically, it considers the fundamental nature of sensory experience, including how individuals conceive and relate to their perceptions (Kenter & O'Connor, 2022). Together, these dimensions emphasize sensory experiences are not universal, instead they are deeply embedded in cultural narratives and individual existential realities, influencing everything from art and language to therapy and personal identity.

Sensory perception is far from monolithic; it fluctuates significantly across cultures, with each culture constructing and interpreting the senses through its own unique lens. This cultural variability profoundly affects how individuals perceive and experience both art and therapeutic interventions. (DeNora, 2014).

The perception of space and aesthetics, molded by cultural, civilizational, and generational influences, crafts the "imageability" of the environment, thereby framing artistic involvement and therapeutic interventions. This highlights the critical need to take cultural context into account when devising sensory experiences in the realms of art and technology (Bhattacharya et al., 2020). Aside from preexisting cultural perspectives, the engagement with art can also promote cognitive and skill development, offering individuals new ways of seeing, feeling, and thinking about their world, thus modifying their experience of it (Song, 2010). The transformative potential of art is particularly pertinent in therapeutic settings, as it allows for a broad range of sensory and cognitive experiences that are culturally resonant and personally meaningful. For instance, in a study comparing sensory analysis of honey between Greek and Chinese panels, it was found that Greek judges used a wider variety of terms, while Chinese judges showed greater consensus agreement for all samples except chestnut, highlighting how cultural backgrounds can influence the perception and description of sensory characteristics (Price et al., 2019). Similarly, culture and culinary habits have been shown to influence edibility ratings and color associations of some odors, with both perceptual and semantic factors playing a role (Tamura & Okamoto, 2023).

Indigenous cultures in the Andes and the Amazon present varied sensory models, with Amazonians giving greater importance to proximity-dependent senses, especially smell (Classen, 1990). This difference underscores the role of environment and cultural practices in shaping sensory priorities and experiences. Cultural factors such as familiarity, health benefits, food labels, language, and the food environment also significantly influence consumer sensory perception and acceptability of foods and drinks, indicating that sensory perception is deeply embedded in cultural context (Jeong & Lee, 2021). Perhaps one of the most interesting phenomena associated with sense perception involves our employment of language. The syntax, structure, and grammar of a language can affect how sensory stimuli are perceived (or not), especially considering color. In other words, one cannot see what one cannot say (Krempel, 2018).

Of the factors influencing sense perception, language significantly affects color perception, with lexical gaps and cultural nuances playing a pivotal role in shaping how individuals interpret colors (He et al., 2019). Research suggests that mapping color is more an active culture than a fact of physics, with relationships and reinforced language creating labeled understandings of what we see (Gage & Grovier, 2023; Galison, 1997). This is exemplified by studies showing that people tend to associate lively music played in a major key with light, warm colors, while low, slow music is associated with hues of blue and gray. As noted, such associations are vaguer manifestations of the literal conjoining of color and sound experienced by people with synesthesia (Hauck et al., 2022). Moreover, languages rich in vocabulary for colors typically have more specific words for shades, influencing the perceptual acuity and description of colors. Italian, Russian, and Hebrew (Fig. 3.2) each has twelve basic color terms and additional words for different shades (Tali, 2022). This allows speakers to be more specific and nuanced in their descriptions of colors, possibly affecting their perception and attention to color variations in their environment. Different cultures and languages conceptualize colors uniquely, influencing perception and categorization. Some limit the number of terms used to describe multiple colors in other languages. For instance, Japanese uses the term *ao* to refer to both green and blue, while Vietnamese uses *hoa* for both red and pink (Song et al., 2023).

The lack of distinction in some languages suggests that cultural context can influence the categorization and perception of colors (Huisman et al., 2022). Likewise, the Dani people of Papua New Guinea, with only two basic color terms, *mili* and *mola,* however, unlike in Japanese or Vietnamese, have an extensive vocabulary for different shades within these categories, including specific terms for colors of various natural elements. The extensive vocabulary allows them to distinguish between shades more accurately than English speakers, suggesting that a rich color lexicon might enhance perceptual discrimination (Enfield, 2022).

Other language systems focus not on the various elements of color—saturation, hue, tine, tone, and shade—instead on other elements. For instance, the Himba of Namibia (Fig. 3.3), with a language system focusing on light absorption and reflection rather than color shades, and the Berinmo people of Papua New Guinea, who distinguish colors based on function or meaning (Fig. 3.4), provide further examples of how

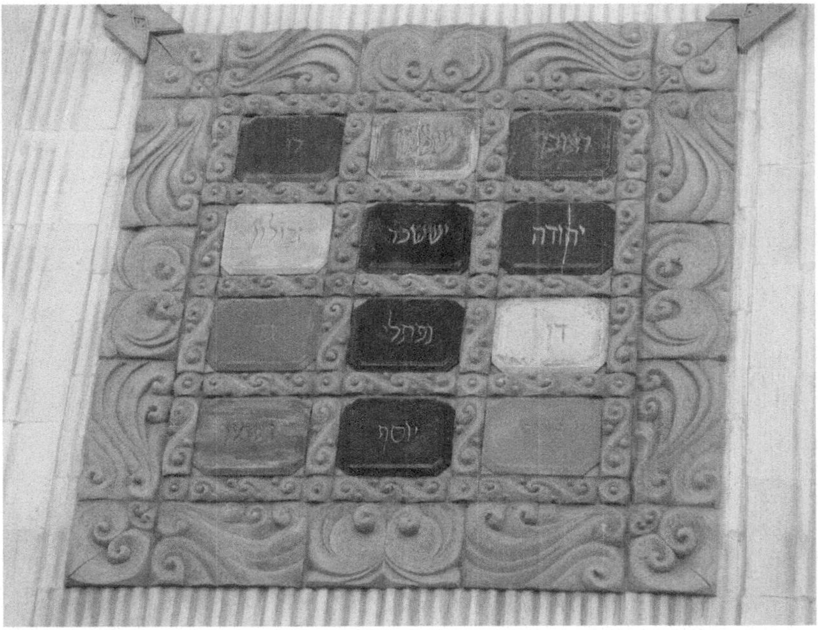

Fig. 3.2 Breastplate on the front of the Central Sephardic, Israel (*Source* From Wikimedia Commons, licensed under CC-BY 2.5)

language shapes color perception (Dedrick, 2022; Mylonas et al., 2022). Hawaiian language, with its extensive vocabulary for colors, textures, and natural phenomena, emphasizes the close link between aesthetics and nature in the culture (Kukahiko et al., 2020). Similarly, the rich vocabulary of the Zapotec for visual phenomena, including shadows, reflections, and light, highlights the intricate relationship between language and perception (Silva-Robles et al., 2022). However, these phenomena exist within the realm of a singular language. The vibrant interplay between light, sound, and language accentuates how sensory experiences are intimately intertwined with linguistic elements, particularly in the context of bilingualism.

The impact of bilingualism on color perception is especially evident, as observed in the indigenous people of lowland Bolivia, the Tsimane'. Here, bilingual individuals who learned Spanish as a second language began to perceive and categorize colors in a way distinct from monolingual Tsimane' speakers. Significantly, these bilingual speakers began to differentiate between blue and green using distinct terms, a linguistic shift that occurred without adopting words from Spanish (Malik-Moraleda et al., 2022). The shift in color perception underscores the profound

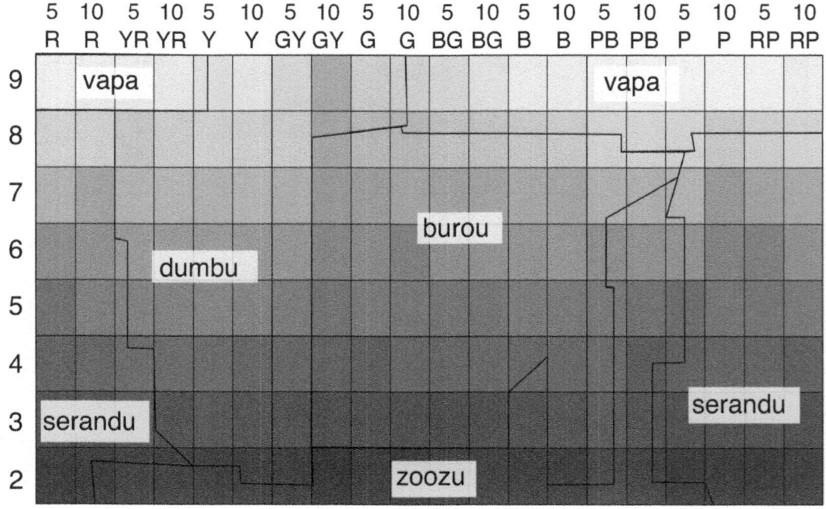

Fig. 3.3 Himba color names (*Source* From Wikimedia Commons, licensed under CC-BY 4)

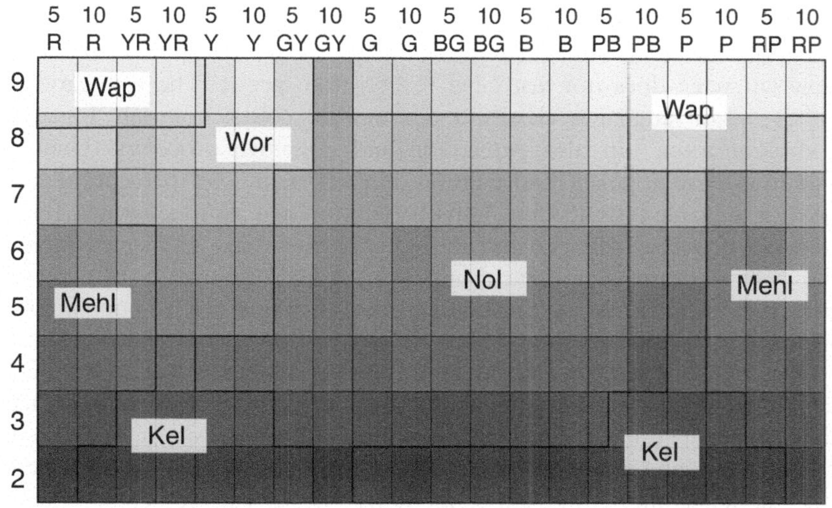

Fig. 3.4 Berinmo color names (*Source* From Wikimedia Commons, licensed under CC-BY 4)

influence of language acquisition on the sensory interpretation of the world, pointing to the broader cognitive implications of bilingualism. On the other hand, polysynthetic languages also offer insight in this area. Further emphasizing the role of language in shaping perception, the study of the Australian Aboriginal language, Murrinhpatha, offers additional insights. The polysynthetic nature of the language allows for the conveyance of complex ideas and actions in single-word expressions. The linguistic architecture of these speakers starkly contrasts with the more stringent grammatical structures found in many Western languages, such as English. The insights gleaned from examining these speakers contest the idea that human world perception is homogenous across different linguistic traditions. Instead, they imply that the distinct linguistic system an individual engages with substantially molds their cognitive functions and perceptions of the world (Mansfield, 2019).

This research highlights the importance of considering language as a significant factor in sensory experiences, given its capacity to modify how individuals engage with and perceive their surroundings. Together, these studies on bilingualism and polysynthetic languages demonstrate how language plays a critical role in shaping sensory experiences and engagement of individuals with their surroundings. These examples underscore how language does not just label reality; it shapes it. The way various cultures and languages classify and articulate colors not only impacts communication, but also perception and cognitive processes. Understanding these linguistic and cultural differences in color perception can provide valuable insights into human cognition, perception, and the relationship between language and thought. As these and further examples illustrate, understanding the cultural and linguistic dimensions of sensory perception is fundamental for training those engaging diverse populations with art, as it allows artists and therapists to develop effective practices and approaches that respect and integrate diverse cultural values, aesthetics, and sensory perceptions. The importance of cultural awareness in art therapy is especially crucial for developing ethical praxis, multicultural competency, and intercultural sensitivity (Kapitan, 2015). Intertwined with understanding the distinct methods various cultures comprehend and convey emotions, symbolism, and meanings through art, these interpretations can differ markedly across cultural boundaries. For example, therapists trained with a strong cultural awareness are better equipped to understand how art is perceived differently across cultures. This might involve recognizing how color symbolism varies across cultural backgrounds or how sensory experiences like taste or smell might be described differently in different languages (Classen, 1993). Such awareness is vital for therapists to avoid misinterpretation of client artwork and to provide interventions that are culturally sensitive and relevant.

Furthermore, training in cultural awareness helps artists and art therapists to become aware of their own cultural biases and assumptions. This self-awareness is essential for creating a safe and supportive therapeutic environment where clients feel understood and valued (George et al., 2005). Addressing culture, values, and aesthetics in art therapy training is therefore not only about understanding client background but also about therapists understanding their own personal and cultural backgrounds, values, and biases (Cattaneo, 1994). In practice, cultural awareness in art therapy might involve incorporating culturally specific symbols, materials, or techniques into therapy sessions, or it might involve therapists educating themselves on the cultural background of the clients they serve. This knowledge can guide therapeutic work with clients, ensuring that interventions are appropriate, respectful, and effective. For example, if working with a client from a culture that values community and collective experience, therapists might incorporate communal art-making activities or consider the role of family in the therapy process (Metzl, 2022). By integrating cultural awareness into art therapy training, therapists can enhance their understanding of and responsiveness to the diverse needs of clients. The same can be said of artists who wish to communicate across cultural differences. This approach not only fosters a more inclusive and empathetic approach for both practitioners but also enriches the therapeutic process and artistic experience, leading to more meaningful and effective outcomes. It is a commitment to understanding, valuing, and incorporating the diverse ways people experience and interact with the world, thereby ushering in possibilities for healing and growth.

Transitioning from exploring cultural dimensions to ontological dimensions, there is a shift from the external influences of culture and society to the intrinsic nature of perception and being. While cultural dimensions provide a context for how sensory experiences are shaped and interpreted externally, ontological dimensions delve into the essence and structure of perception itself (Sabot, 2023). This shift directs attention closer to understanding the fundamental nature of sensory perception, bridging the gap between cultural variability and the core mechanisms that govern sensory experiences and consciousness (Mendelovici, 2023). The ontological dimensions of sensory perception delve into the essence of how individuals experience and interpret sensory information, encompassing a variety of variables and discriminative factors. For instance, sensory discrimination includes dimensions such as quality, intensity, extensity, and duration, which fundamentally shape how individuals perceive their environment (Allen, 2023; Gibson, 1963). These variables are not just passive receptions but are actively constructed by the brain, influenced by both the physical stimuli and the unique physiological and psychological makeup of individuals. For example, culture and inattentional blindness studies show that culture expresses itself distinctly across various sensory modalities due to topological constraints generated by cultural context (Chen et al., 2023; Semin et al., 2023; Wallace, 2007). This suggests that the way individuals perceive or do not perceive sensory information can be deeply influenced by their cultural background, reflecting the complex interplay between sensory perception and cultural ontology. Moreover, cultural context significantly determines both the structure and content of hallucination-like events, linking cultural learning with prediction-based models of perception (Lifshitz & Luhrmann, 2020).

Further exploring the ontological aspects, Cheung et al. (2024) discuss how sensory systems may represent the pleasantness component of perception, suggesting a higher dimensional space of evaluative discriminations that accompany sensory inputs. This indicates that sensory perception is not merely about detecting stimuli but also involves a complex evaluation process that is deeply embedded in individual consciousness and cognitive structures. As such, an ontological structure has been proposed for representing knowledge related to the measure of sensory perceptions, including haptic ones, in smart systems for haptic quality control (Albert et al., 2016). The schema not only underscores the complexity and multidimensionality of sensory perception but also highlights the potential for technological innovations to interface with these ontological dimensions, enhancing our ability to measure, understand, and replicate sensory experiences. This highlights the profound influence of cultural ontology on sensory experiences, shaping not only what is perceived but also how it is interpreted and understood within a cultural framework.

In addition to ontology, intentionalism also offers a compelling perspective in this discourse, proposing how sensory experience is inherently intentional and thus provides a viable ontological and epistemological alternative to traditional theories of sensory experience (Arnaud & Pendoley, 2023; Morvan, 2008). The intentionalist viewpoint spotlights the purposeful and directed nature of sensory perception, suggesting how individuals experience the world is intrinsically linked to their intentions, beliefs, and cultural contexts. In all, these ontological considerations are critical for those engaged in enhancing artistic engagement and therapeutic interventions. They highlight the need for an approach that is not only culturally aware but also deeply cognizant of the varied and nuanced ways in which individuals experience and interpret sensory information. By comprehending and integrating these ontological aspects, practitioners are poised to devise more effective and empathetic strategies. Such approaches respect the diverse tapestry of sensory perceptions informed by cultural and individual variances, culminating in artistic and therapeutic practices with inclusive and impactful potential.

3.2 Aesthetic Disabilities: Beyond the Visual Spectrum and Their Technological Addressal

Building upon their understanding of the ontological considerations of sense perception, this section explores how aesthetic disabilities profoundly influence individual sensory experiences and perceptions. Aesthetic disabilities refer to a range of conditions affecting how individuals perceive, interpret, and appreciate the sensory and aesthetic aspects of their environment, impacting their interaction and experience with the world around them. These conditions include a variety of perceptual challenges, from those affecting the recognition of faces and colors to difficulties in mathematical understanding and the extraordinary blending of senses seen in synesthesia, each presenting unique insights into the diverse ways sensory experiences can be perceived and understood (Table 3.1). As such, this section delves into the complexities of these conditions, examining how they challenge and reshape our conventional understanding of sensory interaction and aesthetic appreciation. It will discuss the intrinsic connection between sensory perception and intentionalism, highlighting how personal and cultural dimensions intersect with these unique sensory experiences (Jackson, 2009). By integrating insights from recent research, the discussion aims to illuminate the ways in which technological interventions can address and enhance the sensory and aesthetic experiences of individuals with these conditions, thereby widening the spectrum of sensory engagement and artistic interpretation.

Delving into the realm of aesthetic disabilities, the discussion first focuses on face blindness, exploring its effects on artistic interpretation and the broader implications it holds for personal and social interactions. Developmental *prosopagnosia*, commonly known as "face blindness," implicates a neural basis in the perception and memory of faces, reflecting the complexity and intentionality in sensory processing (Liu et al., 2021). Research has highlighted the need for a valid taxonomy to facilitate the understanding of the disorder and guide future research (Arizpe et al., 2019). The disorder is linked with bilateral lesions in the central visual system, particularly in the recognition of visually "ambiguous" stimuli requiring evocation of contextual memory, such as faces (Rossion, 2022). Interestingly, studies show unilateral right posterior cerebral lesions might be sufficient to cause prosopagnosia, supporting theories that a right hemispheric lesion can substantially contribute to the condition (Ku & Kim, 2022). Moreover, the right anterior inferior occipital lesions and

Table 3.1 Aesthetic disabilities

Condition	*Description*	*Aesthetic considerations*
Prosopagnosia	Inability to recognize faces despite intact visual capacity	Artists might focus on abstract elements, colors, shapes, and forms rather than facial identity or emotional expressions. Viewers might miss nuances in portraits or expressions
Dyschromatopsia	Difficulty in perceiving colors accurately, often referred to as color blindness	Artists and viewers may have a modified appreciation of color harmony, contrast, and may rely more on texture or form
Dyscalculia	Difficulty in understanding numbers and performing arithmetic tasks	May affect an artist's or viewer's perception of spatial relationships, patterns, and proportions in artworks
Olfactory Synesthesia	A form of synesthesia where certain stimuli trigger olfactory experiences	Artists may incorporate or represent olfactory elements in their work. Viewers might experience art with an added dimension of smell
Lexical-Gustatory Synesthesia	A form of synesthesia where words evoke taste sensations	Can influence creative expression by integrating taste aspects into visual or auditory art, leading to unique artworks that stimulate multiple senses
Anosmia	Loss or reduction of the sense of smell	May influence artists' thematic choices or materials used. Viewers with anosmia may not fully experience artworks designed with olfactory elements

(continued)

Table 3.1 (continued)

Condition	*Description*	*Aesthetic considerations*
Agnosia	Impairment of the ability to recognize and process sensory information	Visual agnosia may lead to unique visual representations by artists. Auditory agnosia affects the perception of sound in art
Aphantasia	Lack of voluntary visual imagination, inability to visualize mental images	May affect artists in their ability to preconceive or imagine scenes or subjects, leading to a different approach to creation and perception of visual arts

often symmetrical left hemisphere lesions at the occipital-temporal junction are associated with prosopagnosia, indicating the anatomical basis of the disorder (Barton et al., 2021). The condition is increasingly understood as a heterogeneous disorder involving impaired face recognition. The mechanism primarily implicated is a content-specific face processing mechanism, suggesting that individuals with prosopagnosia may process faces differently than typical individuals (Avidan & Behrmann, 2021). Some patients show autonomic responses to familiar faces, indicating that early steps of recognition might still be occurring without conscious awareness (Stumps et al., 2020). In addressing the cognitive adaptation, recent research has emphasized the importance of understanding the different functional difficulties across various levels of face processing, as it is not a unitary syndrome (Wilcockson et al., 2020). This diversity in cognitive manifestations supports cognitive models of face recognition and highlights the complexity of developing targeted interventions.

The exploration of prosopagnosia and its impact on individuals provides critical insights into the nature of aesthetic disabilities and underscores the importance of technological and therapeutic interventions to improve the quality of life for those affected by such conditions (Gillespie, 2022). The condition significantly impacts how individuals perceive and interact with art, especially visual aesthetics. Characterized by an inability to recognize faces, it is not just a visual impairment, but also affects the emotional and social experience of viewing art (Freedburg, 2014). One key aspect of prosopagnosia is the impairment in a specialized form of visual recognition necessary for face recognition, which is crucial for interpreting portraits or any art form featuring human faces (Farah et al., 1995). Prosopagnosic patients might experience art differently, focusing more on the abstract elements, colors, shapes, and forms rather than the personal identity or emotional expressions typically conveyed through faces (Leška, 2023).

The contemporary artist Chuck Close (1940–2021), who had developmental prosopagnosia, offers a unique neuroaesthetic perspective as he creates large-scale portraits that challenge and redefine the act of facial recognition (Edkins, 2013). Despite his face blindness, his work, characterized by its enormous scale, featureless backgrounds, and absence of facial expression as in Brad Pitt (2009) (Fig. 3.5), suggests a deliberate artistic strategy rather than a constraint imposed by his condition (https://youtu.be/tXsuo4NWOUY). His approach might be understood as a reflection of his difficulty in holistic processing, common in prosopagnosia, leading to a distinctive artistic style that oscillates between realistic imagery and abstract patterns. This contradiction between his personal struggle with face recognition and his professional focus on creating intricate faces indicates a complex interplay between his neurological condition and artistic expression. The work of Close, by frustrating conventional facial perception and encouraging an alternative, perhaps more map-like or landscape-oriented engagement, invites viewers into a shared space of prosopagnosic experience, highlighting the aesthetic complexity and novel interpretation of familiar faces (Huang & Huang, 2021).

One insight into the work of prosopagnosic artists may be found in research demonstrating patients who show physiological responses to faces they previously recognized but can no longer visually recognize, indicating that early recognition processes might still occur without their awareness (Robotham & Starrfelt, 2018). This suggests the experience of viewing art might still evoke subconscious responses related to face recognition, even in the absence of conscious identification. The loss of "configural processing" in prosopagnosia, which enables immediate identification of individuals without conscious analysis, profoundly impacts how art is perceived (de Gelder et al., 2022). Thus, individuals with the condition, as evinced in the case of Chuck Close, may focus more on individual elements of a piece, such as color and texture, rather than the gestalt of faces, leading to a unique and possibly more analytical

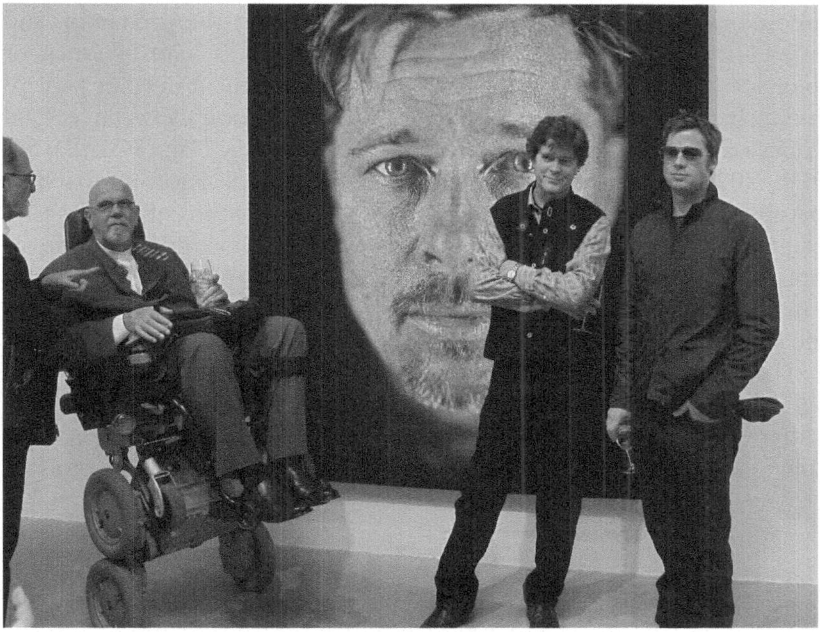

Fig. 3.5 Chuck Close, *Brad Pitt*, 2009 (*Source* From Wikimedia Commons, licensed under CC-BY 3)

approach to art appreciation (Blom et al., 2021). The broader implications of face blindness significantly affect the active capacity for artistic expression and interpretation, potentially leading to unique artistic styles or preferences among affected artists and distinct responses from viewers with the condition. For artists with prosopagnosia, it may influence their choice of subject matter, artistic techniques, or representational strategies, possibly leading to innovative approaches that bypass conventional facial representation. Viewers with this condition might also experience art differently, focusing on aspects other than facial recognition, such as color, form, or the emotional ambiance of artworks. Understanding how prosopagnosia impacts the creation and appreciation of art underscores the need for inclusive and diverse art programs which cater to a wider range of perceptual experiences.

While prosopagnosia challenges individuals with face recognition and leads to different strategies for art viewing and appreciation, other conditions, such as ***dyschromatopsia*** (Fig. 2.3), alter the perception and differentiation of colors. The condition, also known as color blindness or color vision deficiency, is characterized by an impairment in color perception, where individuals find it difficult to distinguish between certain shades of colors or may not perceive colors accurately (Fig. 3.6) (Barbur et al., 2022). To enhance comprehension of the implications of dyschromatopsia, this section intends to explore the most recent studies and insights into the condition. Through investigating how dyschromatopsia influences everyday life, as well as the subtler aspects of art and aesthetics, a more profound understanding of the wide array of sensory difficulties and unique personal experiences may be attained.

The latest research on hereditary ocular diseases has brought forward significant advancements and insights into understanding and potentially addressing this condition (Hassanpoor et al., 2023). Colorblindness, typically manifested as an inability to differentiate between certain colors or

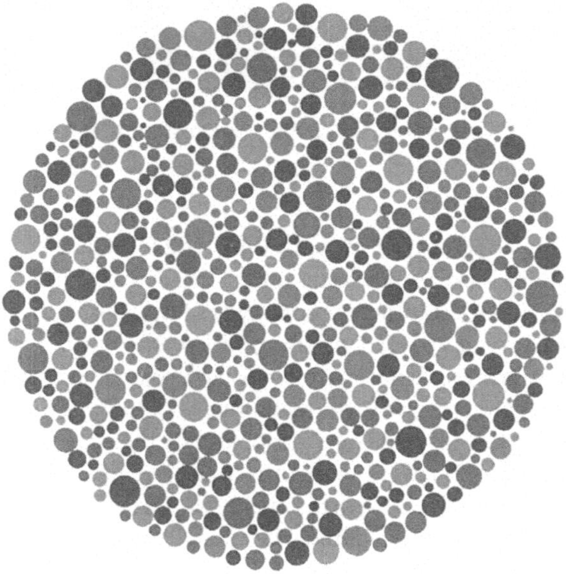

Fig. 3.6 Shinobu Ishihara, colorblindness test (6) (*Source* From Wikimedia Commons, licensed under CC0)

a reduced sensitivity to color, is primarily inherited and affects individuals from birth (Rathee & Mann, 2022). A notable advancement is the exploration of gene therapy as a potential cure for adult red-green color blindness, which seeks to add a third cone pigment to dichromatic retinas, providing trichromatic color vision even without early developmental processes (Sechrest et al., 2023). The approach indicates a promising future where genetic color vision deficiencies might be correctable even later in life. The genetic basis of colorblindness has been elucidated through various studies, such as identifying the achromatopsia locus in the Pingelapese population on 8q21-q22, which is believed to result from population reduction and geographic isolation in the late eighteenth century (Simon et al., 2004). Further, research indicates that mutations in the gene encoding the α-subunit of the cone photoreceptor cGMP-gated cation channel cause total colorblindness, demonstrating a common genetic basis for phototransduction in the human retina (Li, Xu et al., 2023).

Aside from genetic factors leading to the condition from birth, current research also connects the condition to other diagnoses as a comorbidity. In a significant finding, dyschromatopsia in diabetes has been positively correlated with glycosylated hemoglobin levels, indicating that it might be associated with poor metabolic control (Sohrabi et al., 2023). This suggests that monitoring and addressing the condition could be an important aspect of managing diabetes. Further research has shown that a high percentage of diabetics experience colorblindness, particularly when retinopathy is present, highlighting the need for regular screening and intervention to mitigate the impact of the condition (Casciano et al., 2022). Also, dyschromatopsia frequently occurs in multiple sclerosis patients, suggesting an association with worse disease status and potentially subclinical disease progression (Eslami et al., 2022). Furthermore, within education, dyschromatopic subjects show significantly lower general school achievement compared to those without the condition (Gallo et al., 1998). The implications of these findings are manifold. For individuals with dyschromatopsia, the condition impacts everyday tasks and quality of life, including challenges in professions or activities where color discrimination is crucial.

Regarding the impact of colorblindness on artists and art viewers, despite the public familiarity with the condition, it is in fact, a complex and multifaceted issue, altering both the creation and perception of art (Alcaraz Martinez et al., 2022). Notable artists who have experienced color vision difficulties include the famous impressionist Claude Monet (1840–1926), whose cataracts modified his color perception late in life (Fig. 3.7), and printmaker Charles Meryon (1821–1868), known for his etchings of Paris (Fig. 3.8), who was colorblind (Conway, 2022). Their conditions influenced their artistic choices and styles, leading to unique contributions to the world of art. Contemporary creatives like architect Daniel Arsham (1980–), cyborg artist Neil Harbisson (1982–) (Fig. 2.9), and Richie Magbanua (https://www.youtube.com/c/thatcolorblindartist/), among others, continue to navigate and utilize their color vision challenges in innovative ways, questioning traditional approaches to color and composition in art (Manferdini & Benyamin, 2023; Pacolet, 2022). For artists with colorblindness, the condition can lead to a distinct artistic style, often characterized by a focus on form, texture, and contrast rather than traditional color relationships. For instance, some colorblind artists might rely on a more limited palette or employ techniques that emphasize structure and line over color (á Brekku, 2019). This unique perspective can result in visually striking artwork which offers a fresh take on familiar subjects.

Viewers with colorblindness, on the other hand, may experience art differently from those with typical color vision. They might overlook certain nuances or interpret colors in unexpected ways, leading to a

Fig. 3.7 Claude Monet, *Water Lillies (Agapanthus)*, 1915 (*Source* From Wikimedia Commons, licensed under CC0)

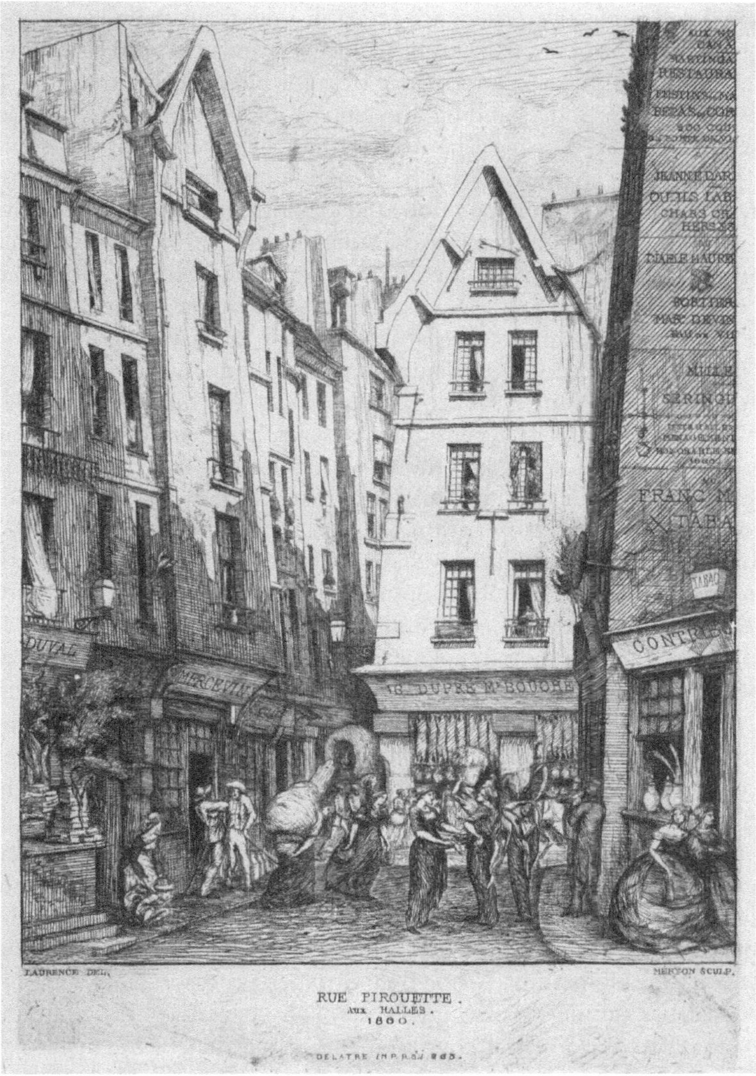

Fig. 3.8 Charles Meryon, *Rue Pirouette aux Halles*, 1860 (*Source* From Wikimedia Commons, licensed under CC0)

personal and subjective experience of art distinct from the general population (Swathi et al., 2020). This difference in perception can also provide valuable insights into the importance of color in conveying meaning and emotion in art. Understanding the impact of colorblindness on both artists and viewers is crucial in appreciating the diversity of experiences and perspectives in the art world. It highlights the need for inclusive strategies in art education, curation, and exhibition, ensuring that individuals with color vision difficulties can fully engage with and enjoy art. By embracing these differences, the art community can foster a more inclusive environment that values and celebrates the unique contributions of all artists and art enthusiasts, regardless of their sensory experiences.

The sensory processing challenges in colorblindness have been extensively researched, but other cognitive processing difficulties, such as with ***dyscalculia***, offer other insights. Dyscalculia, a neurodevelopmental disorder impacting 4–6% of the US population often called "math dyslexia," is a specific learning disability that affects the ability to understand numbers and learn arithmetic facts (Hussain & Soares, 2022). Individuals with dyscalculia may have difficulties with various numerical tasks, including basic number sense, counting, calculation, and understanding mathematical concepts and symbols (Decarli et al., 2023). The condition is intrinsic and often appears early in life, affecting daily life and academic performance of those affected (Santos et al., 2022). Children with developmental dyscalculia often show reduced gray and white matter volumes in the frontoparietal network, which can impact their arithmetic processing skills, fact retrieval, and spatial memory processing (Michels et al., 2022). These neurological findings are crucial for understanding the neurocognitive underpinnings of dyscalculia and developing targeted educational and therapeutic strategies. Recent studies have broadened the understanding of dyscalculia, emphasizing its impact on various cognitive processes (Agostini et al., 2022). For instance, developmental dyscalculia in adults impairs not only number processing but also non-numerical magnitude processing and domain-general skills, highlighting the pervasive nature of the condition (Göbel et al., 2022). In terms of interventions, recent advancements include the use of interactive video media and mobile applications to improve mathematical learning and cognition in individuals with dyscalculia (Dehghani, 2019; Widodo et al., 2021). These technologies offer innovative ways to address learning challenges, providing dyscalculic learners with tools to enhance their mathematical abilities.

These neurological insights have significant implications for how individuals with the condition interact with and interpret mathematical elements in art, particularly in terms of how artists and viewers perceive and incorporate mathematical constructs into their artistic endeavors. The connection between the visual arts and mathematics has a storied history. No less than the Renaissance artists and polymath Leonardo da Vinci (1452–1519) wrote: "Let no one read me who is not a mathematician" (Burke, 2020). But what about artists who cannot engage in mathematics? Artists with dyscalculia may approach their work with a unique perspective, focusing less on geometric precision or symmetry and more on intuitive or abstract representations (Deh, 2019). Their perception of space, proportion, and numerical relationships in art can lead to innovative styles and methods, producing works that emphasize other elements like texture, narrative, or emotion. For instance, Margot Connop (2005–) (https://dyscalculia-blog.com/2022/03/23/artist-margot-connop-art-dyscalculia/), an artist with dyscalculia, uses her condition as a source of unique artistic inspiration, focusing on how humans perceive the world through her video art and moving images. Diagnosed at age 9 along with dyslexia and dyspraxia, Connop's understanding of her dyscalculia has allowed her to approach her difficulties as part of her unique identity, using art as a medium to express and navigate her world. She embraces the lack of definitive answers in the visual arts as a liberating contrast to the stringent structure of mathematics, providing a space for freedom and personal growth.

Viewers with dyscalculia, on the other hand, might experience art differently from those without the condition. They may be more attuned to aspects of the artwork that are not directly related to mathematical precision or symmetry, such as color, emotion, and general composition. They might also perceive abstract or non-representational art in a unique light, finding meaning and beauty in the patterns and forms that are not bound by conventional mathematical aesthetics. Moreover, research indicates that dyscalculia involves impairments in areas dedicated to number magnitude processing and, to some extent, domain-general magnitude processing (Klein & Knops, 2023). This means that both creating and appreciating art that involves numerical concepts or spatial relationships could be challenging for individuals with dyscalculia, as their perception and processing of these elements are affected.

Acknowledging the impact of dyscalculia on the creation and enjoyment of art is crucial for fostering an artistic community that is inclusive and understanding. This awareness calls for the creation of innovative educational and appreciative methods in the arts that cater to diverse cognitive experiences. Such approaches will enhance the expression and appreciation of art in its many forms. Embracing the unique perspectives of individuals with dyscalculia allows the art community to access a wider range of insights and viewpoints, enriching the cultural and aesthetic landscape universally. As we acknowledge the unique ways in which dyscalculia impacts engagement with the mathematical aspects of art, we transition to considering other sensory intersections, such as olfactory and lexical-gustatory synesthesia, where the blending of senses further diversifies the experience and interpretation of art. This shift highlights the broad spectrum of sensory and cognitive experiences which deepen our comprehension and enjoyment of the world surrounding us.

The phenomena of *olfactory and lexical-gustatory synesthesia* offer a fascinating window into how some individuals experience stimuli in uniquely multisensory ways. Olfactory synesthesia is a rare form of the condition where certain information, such as words or sounds, trigger olfactory (smell) experiences. Lexical-gustatory synesthesia is even more specific, involving taste sensations when reading, hearing, or imagining words (Fretwell & Hsu, 2023). These forms of synesthesia involve the crossing of sensory pathways, leading to a variety of experiences such as smelling odors or tasting flavors in response to specific words or sounds (Yan et al., 2021). Research into olfactory synesthesia found individuals with the condition can experience strong and automatic olfactory percepts when presented with congruent visual images (Ward et al., 2020). This suggests their experiences may be more than just vivid imagery, indicating a genuine sensory crossover. Similarly, lexical-gustatory synesthesia is characterized by experiencing taste sensations when reading, hearing, or imagining words. This condition is associated with emotional valence and activation in the insula and medial parietal lobe, reflecting the emotional and physiological nature of these experiences (Ipser et al., 2020).

Olfactory and lexical-gustatory synesthesia profoundly reshape how individuals perceive and engage with art, leading to a complex interplay of sensory experiences. These conditions are rooted in neural activation patterns such as insula and medial parietal lobe activity, and therefore are deeply emotional and physiological in nature (Riedel et al., 2020). Artists with these types of synesthesia may produce work transcending visual or auditory elements to embody taste or smell, thereby offering a more immersive and sensory-rich experience (Wu & Gingrich, 2020).

For instance, those with lexical-gustatory synesthesia might incorporate specific tastes evoked by words into their creative process, while those with olfactory synesthesia may emphasize smell in their artwork. American artist and filmmaker Lucy Engelman (1987–) is such an example as she can taste colors, smell sounds, and even perceive verbal data as a vibrant spectrum of colors. With the intention to convey the complexities of her sensory world to those without synesthesia, her husband, Scottish painter Daniel Mullen (1985–), embarked on a collaborative project titled *A Different Kind of Time: Sequencing Spatial Temporal Synesthesia* (https://danielmullen.info/). This series of paintings features sequences of flat rectangular shapes arranged in various arches and lines, with the orientation of the shapes shifting across each artwork. Some appear to leap off the canvas, while others recede into the rotated plane of the painting. According to Engelman, these works are the closest visual approximation of her synesthetic experiences, particularly how her mind translates letters, numbers, and time (Stevenson, 2019). Through their artistic collaboration, Engelman and Mullen bridge the gap between the synesthetic experience and the typical sensory perception, offering a glimpse into the richly layered world of synesthesia (Sierzputowski, 2017).

The unique sensory entanglement deeply influences their artistic expression and the perception of the viewer, expanding traditional notions of aesthetics and sensory perception. Through their creations, these artists encourage audiences to immerse themselves in a multidimensional world, diminishing the distinctions among the senses and augmenting the art experience (Loh, 2023). Conversely, viewers might perceive art as evolving into a profoundly multisensory voyage, where visual or auditory stimuli can concurrently elicit smells or tastes. Such a sensory confluence can significantly deepen and individualize the art experience, rendering it more emotionally profound and enveloping. As sensory boundaries blur, each artwork becomes a unique exploration, potentially triggering a distinct combination of olfactory or gustatory sensations (Rudenko & Haahr, 2023). The enhanced sensory engagement allows for a more profound and individualized interaction with art, as viewers navigate and appreciate the complex dimensions of sensory input each piece offers.

The sensory intersections encountered through olfactory and lexical-gustatory synesthesia permit a broader array of sensory experiences. However, what might the implications be if these very senses experienced difficulties in processing stimuli? For example, unlike the multisensory experiences of synesthesia, *anosmia* represents a sensory deficit, often leading to a significant impact on quality of life and experiences, including interactions with art and the environment. Anosmia is a medical condition characterized by the loss of the sense of smell (Saltagi et al., 2021). It is crucial to distinguish that "nose blindness" differs from anosmia, which is characterized by a significant reduction or complete loss of the sense of smell. Nose blindness typically occurs as a natural, gradual adaptation, whereas anosmia can result from a variety of causes including nasal congestion, head injury, infections, or can be a symptom of neurological conditions (Klopfenstein et al., 2020). The condition can significantly affect quality of life, impacting taste and food enjoyment, personal safety, and overall well-being (Saniasiaya & Prepageran, 2021). Recent research highlights the prevalence and importance of the condition, especially given the COVID-19 pandemic, where it is a key symptom. For instance, 1 in 6 patients with recent onset anosmia report it as an isolated symptom, indicating its potential role in identifying asymptomatic carriers of the disease (Hopkins et al., 2020). Moreover, anosmia is often associated with a variety of other conditions, including traumatic brain injury (TBI), showing a high incidence of anosmia with increased severity of injury and a significant link to clinical depression (Laing et al., 2021).

Anosmia, the absence of the ability to perceive odors, carries substantial consequences for both the creators and appreciators of art. Originating from congenital factors, head injuries, or various other reasons, this condition drastically changes the sensory world of individuals. It impacts not only their ability to detect scents but also deeply affects their emotional and aesthetic responses (Langdon et al., 2021). For artists with anosmia, the absence of olfactory cues can influence the themes, materials, and sensory focus of their art (Finlay, 2020). They might lean toward more visually or texturally rich mediums, or they might explore the concept of absence and sensory compensation in their work. For individuals with anosmia, particularly those with congenital anosmia or long-lasting acquired anosmia, there have been compensatory abilities reported in other sensory domains, such as an increased ability to detect emotions from facial expressions (Lemogne et al., 2015). The adaptation could lead to unique interpretations and appreciations of art, focusing more on the emotional content or visual cues present in artworks. The loss of smell can also affect the quality of life, including the enjoyment of food and the detection of environmental hazards, which can indirectly influence mood and thematic exploration of artists (Boesveldt et al., 2017). The psychological impact of anosmia, including feelings of isolation and emotional blunting, can become themes in art, reflecting the deeper existential and sensory inquiries that artists with anosmia might explore (Schablitzky & Pause, 2014).

Dubai-born multimedia artist based in New York City, Adnan Aga, was born anosmic, and has creatively confronted his inability to perceive scents by developing Adnose, an experimental prosthetic that employs Computer Vision (CV) and Artificial Intelligence (AI) to hypothesize the smells of various objects. The inventive device is not only a personal aid for Aga but stands as a representation and extension for the broader anosmic community, reflecting their shared curiosity about the scented world around them. The presence of the device in exhibitions, particularly at events like Smell-and-Tell at the Olfactory Art Keller (https://www.olfactoryartkeller.com/scent-performance-series/adnose), allows visitors to engage with the device, testing it with a variety of objects to create smell descriptions. Through this interactive art, Aga draws attention to the nuances of olfactory perception and the unique challenges faced by individuals with anosmia, pushing the boundaries of sensory art and highlighting the transformative potential of technological innovation in bridging sensory gaps. His work serves as a poignant reminder of the often overlooked sense of smell and its profound impact on personal and shared experiences.

For individuals with anosmia, engaging with art might omit the olfactory dimension, a loss especially felt in artworks intended to provoke a multisensory response or within environments such as gardens or installations crafted with scent elements (Spence, 2020). However, the absence of smell might also heighten other sensory modalities or lead to a more imaginative engagement with art, as the brain compensates for the missing sense. Thus, the condition profoundly impacts both the creation and appreciation of art, highlighting the importance of a multisensory approach in understanding aesthetic experiences. It challenges artists and viewers alike to consider the full spectrum of sensory inputs and compensations, enriching the dialogue around sensory experiences and art. And while the loss of one sense, such as that of smell, has a demonstrable impact on sense experience with art, other conditions impact a broader spectrum of external stimuli and their processing.

Anosmia, loss of the sense of smell, finds its counterpart in the loss or impairment of the ability to recognize stimuli through other senses, known as *agnosia* (De Renzi, 2020). The condition can manifest in various forms, but two notable types are visual and auditory agnosia. Visual agnosia is the inability to recognize visually presented objects despite having intact vision, while auditory agnosia refers to the inability to recognize or differentiate sounds despite having functional hearing (Miceli & Caccia, 2023). These conditions indicate a disconnect between sensory input and the ability of the brain to interpret or give meaning to that input, despite often normal functioning of the sensory organs themselves (Bartolomeo & Miceli, 2023). Research on these conditions has provided further insights into their biological and neurological underpinnings. For instance, auditory agnosia is often linked with repeated temporal lobe damage, affecting modality-specific deficits and fluctuating performance (Miceli & Caccia, 2022). Visual agnosia, on the other hand, may develop acutely as the sole disability in individuals with high intelligence and good personality, indicating significant neurological implications (Peel & Chouinard, 2023). These conditions suggest a modular architecture for sensory processing, where distinct pathways and cortical areas are responsible for specific sensory modalities (Stokes, 2023).

Within the community of artists and their audience, visual and auditory agnosia introduce distinct challenges, profoundly changing the dynamics of artistic creation and interpretation. Artists suffering from visual agnosia, defined by the inability to recognize objects seen despite having unimpaired vision, may struggle with deciphering complex objects or scenarios, which can influence their capacity to represent subjects precisely (Cárdenas-Belaunzarán & Cerrillo-Avila, 2023). Conversely, this could also lead to a unique visual vocabulary in their art, as they might focus more on individual elements, abstract forms, or the overall emotional impression of a scene (Chatterjee, 2006). Their creations might reflect a distinctive interpretation of the world, with a focus on alternative sensory details or conceptual elements. Such was the case of a 73-year-old left-handed male artist with visual agnosia following an occipital cerebral vascular accident provides a compelling example of how neurological disorders can impact artistic expression. As Wapner et al. (1978) reported, despite his general inability to identify single objects visually and difficulty interpreting complex scenes, the artist retained the capacity to recognize geometric forms and perceive optical illusions, as well as to accurately copy designs and objects, indicative of visual agnosia. Remarkably, while he could not recognize objects or scenes, he maintained various artistic techniques such as perspective, shadowing, and texture designation, enabling him to copy displays veridically. Surprisingly, when displays were recognized or when he was prompted with just the name of an object to draw, he exhibited a more creative and less literal approach, reminiscent of his pre-morbid artwork. The comparison of his post-morbid and earlier works reveals the nuanced impact of his condition, showing both a strong similarity to his former style and signs of the effects of the condition, such as less distinction between figure and ground, overelaboration, and areas of neglect. His adapted strategies, involving compensatory verbal and motor procedures, underscore the complex interplay between cognitive impairment and artistic creativity, offering a unique window into the resilience and adaptability of the artistic process in the face of neurological challenges.

Aside from physiological agnosia, critics have also discussed *aesthetic agnosia*. In the case of the contemporary artist Jeff Koons (1955–) offers an example where the embodiment of the term, akin to associative agnosia, finds him recognizing and describing objects but failing to understand their significance or context. In the case of this American sculptor, the condition is not neurological but an artistic stance where he deliberately refuses to differentiate between the cultural and emotional values of high art and low art, merging them in his works (Gopnik, 2017). His art, famous for blending high and low culture—like tulips (Fig. 3.9) turned into monumental sculptures and porcelain renditions of Michael Jackson—challenges traditional aesthetic hierarchies and meanings. The approach to art taken by the artist, devoid of irony or guile, treats all subjects, whether classical sculptures or pornographic images, with the same level of importance and inquiry. This intentional blurring of boundaries leads to fresh, significant works that still resonate after decades. Through his art, Koons advocates for a broad, inclusive view of aesthetic experience, urging viewers to accept and appreciate all aspects of life and art without prejudice. His work represents a refusal to adhere to conventional aesthetic categories, suggesting a sort of aesthetic agnosia as a deliberate artistic strategy rather than a cognitive impairment, expanding the horizons of what can be appreciated and understood in the art world (Koons, 2014).

Inversely, auditory agnosia is characterized by the incapacity to identify or distinguish sounds, altering both artists and observers in the domain of auditory art, including music or sound installations. Those with auditory agnosia may find themselves leaning on visual cues or alternative sensory inputs when engaging with auditory art. Alternatively, they might

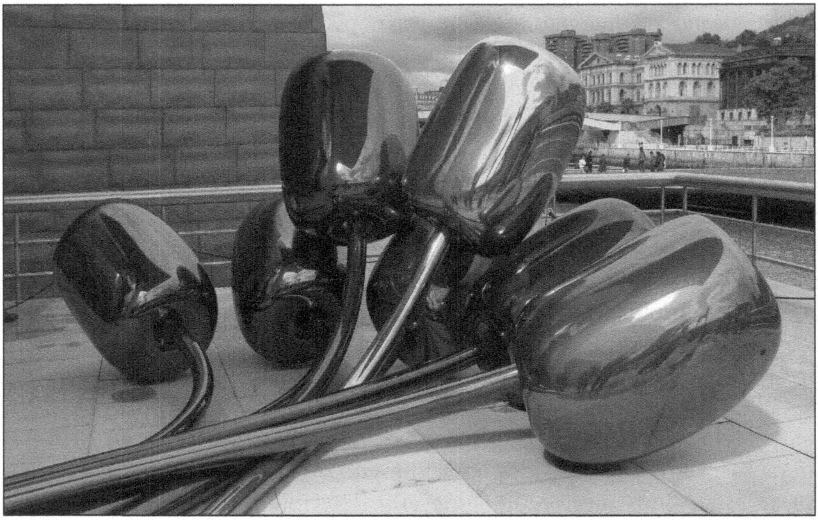

Fig. 3.9 Jeff Koons, *Tulips*, 1997. Guggenheim Bilbao, Spain (*Source* From Wikimedia Commons, licensed under CC-BY 2)

perceive sounds in a fundamentally altered manner, resulting in distinct interpretations and experiences of auditory art, potentially leading to novel creations or encounters (Miceli & Caccia, 2023). For viewers with visual agnosia, the experience of art often lacks the immediate recognition of objects or scenes, leading them to engage more deeply with the color, texture, or emotional content of the artwork (Ptak & Lazeyras, 2019). Similarly, viewers with auditory agnosia might focus on the visual aspects of a performance or the physical vibrations of sound, finding alternative ways to connect with the auditory art (Watanabe et al., 2020). Understanding the impacts of visual and auditory agnosia on artists and art viewers is crucial for appreciating the diverse ways individuals create and experience art. It asserts the importance of accommodating different sensory experiences in art education, exhibition, and appreciation, ensuring that everyone can engage with and enjoy art, regardless of their sensory capabilities. But what if an exhibition or museum visitor lacks the ability to visualize mental imagery at all?

Aphantasia, distinct from agnosia and the inability to recognize familiar stimuli, is a neurological condition defined by the absence of voluntary visual imagination, leading to a pervasive lack of mental imagery, particularly in the visual domain (Arcangeli, 2023). The phenomenon affects how individuals mentally visualize scenes, objects, or concepts, and extends beyond just visual aspects, touching upon multiple sensory modalities and shaping a multisensory experience. Individuals with aphantasia lack a functioning "mind's eye," meaning they cannot conjure up images, scenes, or pictures in their imagination. Originally described by Francis Galton in 1880 and later defined in a 2015 study by Professor Adam Zeman and his team at the University of Exeter, aphantasia varies in its presentation, with some individuals experiencing minimal to no visual imagery and others involuntary mental imagery (Fox-Muraton, 2021). Despite the absence of a cure and the rarity of the condition, increased awareness and research are shedding light on its underlying mechanisms and potential approaches to understanding and living with the neurological condition (Keogh et al., 2021).

Recent research into aphantasia shed light on its various facets and implications (Muraki et al., 2023). Individuals with congenital aphantasia show no significant neuropsychological deficits on imagery-related memory tasks but may perform slower in tasks with increased working memory load, such as the One Touch Stocking of Cambridge task (Aslan et al., 2023). These findings indicate while the condition affects the ability to voluntarily generate mental images, it does not necessarily impair other cognitive functions related to memory or imagination. Aphantasics might also experience abnormalities in cortical activity involved in visual perception and imagery, particularly in the left temporal areas that are typically involved in evoking mental images (Furman et al., 2022). This abnormal activity could be indicative of the underlying neurological basis of the condition and its role in visual working memory and attentional guidance. Another study quantified the phenomenon through drawing tasks, finding that aphantasics show object-specific memory impairment but retain high spatial accuracy (Fan et al., 2023). This suggests though individuals with aphantasia may struggle to recall or visualize specific objects, their understanding and memory of spatial relationships remain intact.

Awareness of aphantasia is crucial for comprehending how the absence of mental imagery plays into various aspects of life, including creativity, memory, and day-to-day functioning (Tosserams et al., 2022). The cognitive implications of aphantasia challenge traditional art experiences for both creators and viewers, reshaping their engagement with art. For artists with aphantasia, the inability to form and work with mental images alters their consciousness engagement with the image-object, which may question aspects of an existential account of self (Fox-Muraton, 2021). Instead, relying on external structures and physical manipulations of external media to generate artwork and demonstrating that visual imagining can extend beyond the brain-bound "hylomorphic" model (MacKisack et al., 2020). This infers that aphantasic artists adapt their creative processes, often relying on direct observation, conceptual thinking, or other sensory inputs, leading to distinctive artistic styles and/or methods.

Glen Keane (1954–) is celebrated as one of the greatest animators in hand-drawn animation and exemplifies the intriguing capabilities of individuals with aphantasia in the creative arts (Hart & Hay, 2022). Despite his inability to generate mental imagery, Keane has been responsible for some of the most iconic characters in animation, such as Ariel from *The Little Mermaid* (1989) and the Beast from *Beauty and the Beast* (1991). The absence of preconceived visual imagery in his mind does not hinder his creative process. Instead, the animator relies on a deep knowledge of anatomy and form, combined with a dynamic and iterative drawing process, to bring characters to life (Keane, 2021). His approach involves starting with an "explosion of scribbles," progressively refining and subtracting lines to discover the desired form. This method, underpinned by a feedback loop of perception and modification, allows him to create detailed and expressive characters without needing to visualize them mentally. His work with *the Beast,* incorporating features from various animals and fine-tuning until achieving the right expression, demonstrates how knowledge, perception, and physical interaction with the medium can compensate for the absence of mental imagery, leading to profound artistic creation. In such a way, artists with this condition claim to be able to remember details of things seen, yet they are unable to visualize them as complete entities (Bracey, 2020).

Conversely, individuals with aphantasia engage with art in a distinct manner, characterized by diminished mental imagery in various sensory realms, less vivid recollections, and impoverished dreams (Dawes et al., 2020). This results in more concrete and immediate engagements with artworks, focusing on the physical and conceptual qualities present rather than relying on mental imagery to fill in or enhance the experience. However, their ability to perform equally well in tasks akin to change detection supports the "overflow" account of perceptual consciousness, suggesting while their internal visual world might be less vivid, their perception of the external world remains rich and detailed (D'Aloisio-Montilla, 2017). The impact of aphantasia on art experience is significant, necessitating a reconsideration of how art is taught, created, and appreciated. By understanding the unique challenges and perspectives brought by aphantasia, artists, educators, and viewers can foster more inclusive and diverse approaches to art, accommodating a wider range of cognitive experiences and enriching the cultural and aesthetic landscape. As research continues, it will be important to explore further the ways in which aphantasia affects the artistic process and the art experience, guiding more empathetic and innovative approaches to creativity and appreciation.

In concluding this exploration of aesthetic challenges, it is clear such conditions like prosopagnosia profoundly reshape sensory perception and artistic expression. These discussions illuminate the rich complexity of human experience and the importance of accommodating diverse sensory perspectives in art and education. As the narrative shifts from the intricacies of individual sensory experiences, the next theme emerges: the role of technology in addressing and ameliorating cross-modal sensory challenges. The next section will delve into how innovative solutions can bridge the gaps in perception, offering new avenues for individuals to engage with their environment and expand the boundaries of their sensory world. This shift signifies a transition in focus and a continuation of the commitment to understanding and enhancing human interaction in a multisensory context.

3.3 Neurodivergent Sensory Preferences and Technological Accommodations

Along with understanding individuals with diverse aesthetic adaptations, it is equally important to consider the sensory preferences and corresponding technological accommodations for neurodivergent populations, including those on the spectrum, ADHD, PTSD, and related conditions. Neurodivergent individuals frequently experience the world in unique ways, characterized by distinct sensory preferences and challenges (Morgan, 2019). These experiences necessitate thoughtful accommodation and the development of innovative solutions to meet their varied sensory needs effectively. Accordingly, recent research emphasizes the importance of designing sensory-sensitive technologies that minimize discomfort and maximize usability for neurodivergent individuals. For example, understanding design preferences for sensory-sensitive earcons with neurodivergent individuals reveals a preference for designs that follow six key themes for effective design (Race et al., 2022). Additionally, neurodivergent user experience of XR technologies highlights the struggle with sensory overload and the need for customized sensory settings to make these technologies more accessible (Hutson & McGinley, 2023).

As will be elaborated in Chapter 4, immersive reality technologies afford the ability to modulate and even ameliorate certain sensory conditions (Fig. 3.10). Extended Reality (XR), encompassing Augmented Reality (AR), Mixed Reality (MR), and Virtual Reality (VR), offers a versatile spectrum known as the reality continuum that effectively addresses various sensory processing challenges. By blending real and virtual worlds, XR technologies provide immersive experiences which can be tailored to individual sensory needs. For instance, VR can create controlled visual environments for individuals with visual processing challenges, allowing them to interact with and adapt to various scenarios in a safe space. Similarly, AR overlays digital information onto the real world, aiding those with auditory challenges by providing visual cues or subtitles in real time. MR combines aspects of both VR and AR, creating interactive environments where physical and digital objects coexist, which can be particularly beneficial for tactile learning and therapy, providing sensory feedback and enhancing motor skills. Furthermore, advancements in olfactory technology are integrating scent into XR experiences, offering therapeutic possibilities for individuals with olfactory processing issues. These immersive technologies are revolutionizing the way sensory processing challenges are approached, providing customizable and adaptive solutions that cater to a wide range of sensory needs, thereby enhancing learning, therapy, and the overall quality of life for individuals with sensory processing disorders. Likewise, in the realm of education, neurodivergent students, particularly those with ADHD, benefit from emerging technologies that cater to their sensory, control, and realism needs while addressing the challenge of maintaining attention (Hite et al., 2021). Furthermore, sensory-friendly environments have been identified as crucial for increasing the quality of life and accessibility for the neurodivergent autistic minority, especially those with hypersensitivities (Morgan, 2019).

With such considerations in mind, this section investigates the latest research and developments in understanding and accommodating neurodivergent sensory preferences, examining how technological innovations can be harnessed to create more inclusive and effective environments for

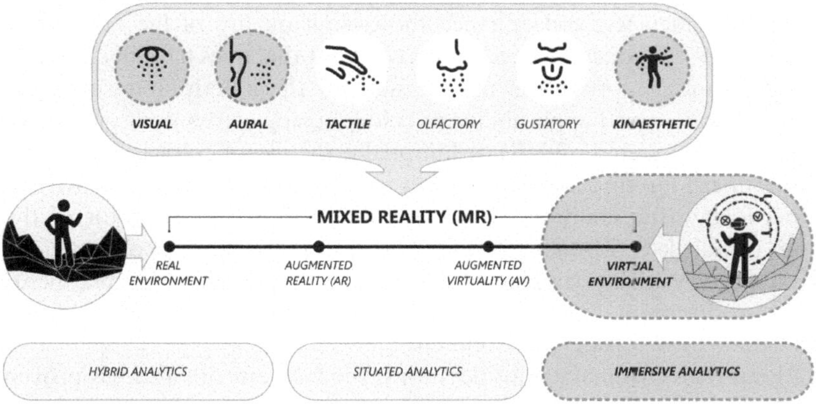

Fig. 3.10 Reality continuum. From Moloney, Spehar, Globa, and Wang (2018) (*Source* From Wikimedia Commons, licensed under CC-BY 4)

learning, working, and living. One of the more well-known interventions acknowledging the diverse sensory worlds of neurodivergent individuals is *color therapy*. Also known as chromotherapy, it posits certain colors can have beneficial effects on the nervous system and brain function, making it a potentially helpful approach for individuals with neurodivergence, by reducing sensory overload and aiding in creating more comfortable environments (Cheron et al., 2022). Fluorescent lights, in particular, are believed to be overstimulating for people with sensory sensitivities, and the flickering and harsh colors can trigger sensory overload (Martin & Wilkins, 2022). The use of specific colors in therapy aims to balance this effect and provide a more soothing experience (Wei et al., 2023).

Research in color therapy suggests its potential as a non-invasive treatment option for various medical and dental diseases (Sindhuja et al., 2022). Brain photobiomodulation therapy using red to near-infrared light has shown promise in enhancing neuron metabolic capacity and stimulating anti-inflammatory, anti-apoptotic, and antioxidant responses, which can be beneficial for neurological and psychological conditions (Salehpour et al., 2018). Furthermore, color therapy has been found to effectively reduce stress and anxiety levels, although it is noted that factors other than physical exposure to color play a role in these changes (Jonauskaite et al., 2023). In the context of neurodivergent conditions, color therapy can serve as a low-cost, non-invasive option which may improve mood, reduce hyperactivity, and enhance the overall quality of life (Dorriy & Shool, 2019). For instance, red and green light therapies have been shown to significantly improve the mental state of the elderly, indicating the potential widespread applicability of color therapy across multiple demographics (Chang et al., 2018). Additionally, the use of color in playrooms with artificial lighting has been shown to effectively increase sensitivity to colors for autistic children without changing the shape or color of the room (Habbak & Khodeir, 2023). This research indicates color therapy may offer a valuable supplementary approach to traditional treatments for neurodivergent individuals, helping to create a more balanced and accommodating sensory environment.

Along with chromatic considerations, the tactile realm also has proven important for neurodivergent populations. For example, the role of smooth textures in artistic experiences, particularly for individuals on the spectrum who may have a heightened affinity for such textures, presents intriguing implications for both artistic creation and perception. Smooth textures are often preferred for their tactile qualities and can influence aesthetic appreciation and sensory comfort (Cascio et al., 2012). This preference for smoother or softer stimuli over rougher or harder textures can significantly affect how individuals with autism and other neurodivergent conditions interact with art and other tactile stimuli. On the other hand, for artists with a preference for smooth textures, this can lead to a unique approach to material selection and artistic expression, often incorporating these textures into their work to create art that is not only visually appealing but also tactilely engaging. The smoothness of texture has cross-modal correspondences with olfactory stimuli, which may influence how artists conceive and viewers perceive art, with emotions and knowledge of an odor's identity mediating these effects (losifyan et al., 2022).

Viewers with neurodivergence, particularly those who may have heightened or diminished sensitivity to sensory inputs, can experience a more profound connection to artworks featuring preferred textures. The tactile quality of an artwork can evoke a strong emotional and aesthetic response, making the experience of viewing art more immersive and personal. The preference for smooth textures over rougher ones is evident in tactile preferences and aesthetic appreciation, indicating that the physical texture of materials can have a profound impact on how art is experienced (Karim et al., 2022). Moreover, the sensory experience of texture in art extends beyond the tactile to the visual domain, with the smoothness of surfaces potentially influencing the visual perception of artworks. For individuals with sensory processing differences, the textural qualities of art can play a critical role in their engagement and enjoyment of art. As such, understanding textural affinities and preferences is crucial for creating inclusive and accessible art experiences catering to neurodivergent individuals (Elmarakbi et al., 2023).

In further exploring aesthetic preferences, particularly among those outside of neurotypical perceptions, color and texture are but two considerations. Detail recognition and appreciation are also often anchored in neurotypes, leading to unique experiences and interpretations of art. Neurodivergent individuals tend to have enhanced reliance on visual details, which can significantly impact their aesthetic preferences and how they interact with various forms of art (Alink & Charest, 2020). These tendencies highlight a more detail-oriented visual information processing strategy, potentially generalizing to a clinical ASC population and affecting various real-life abilities and difficulties. Research utilizing deep convolutional neural networks trained for object recognition has further demonstrated that human preferences for art can be predicted from a mixture of low- and high-level visual features (Iigaya et al., 2021). This suggests that aesthetic preferences are not just subjective but can be systematically understood and anticipated based on visual details and structures. Furthermore, individual differences in aesthetic preferences, such as preferences for harmony across various domains, including color and shape, have been shown to correlate strongly and decrease with training in art and music (Elkoshi, 2019).

The connection between aesthetic preference and detail involves not just music and art, but also three-dimensional objects. EEG studies on 3D shapes also reveal that aesthetic preferences can be recognized with up to 80 percent accuracy, indicating that brain responses to aesthetic stimuli are somewhat consistent and predictable (Suhaimi et al., 2020). Similarly, aesthetic preference for paintings has been associated with specific brain structures, which play a role in evaluating reward-based stimuli with different emotional valence (Iigaya et al., 2021). The findings tie into the broader understanding of neuroaesthetics and how the reception and processing of details by the brain can shape aesthetic experiences, which also extend to an understanding of the nature of reality as perceived by individuals (Derke et al., 2023).

The concept of "presence-continuum confusion" or "virtual reality confusion," especially prevalent in individuals with autism, represents a significant challenge as it pertains to distinguishing between the physical and virtual worlds (Bouchard & Rizzo, 2019). The confusion is theorized to occur because the rich and immersive sensory information from the virtual world overwhelms the ability of the brain to differentiate it from the physical world. Such difficulty in distinguishing between virtual and real stimuli is particularly pertinent in ASC, where sensory processing can already be atypical (Neufeld et al., 2021). The interoception sensory questionnaire reveals a significant proportion of autistic adults report interoceptive confusion, affecting their physiological self-regulation and physical health outcomes (Fiene et al., 2018). Moreover, neurological soft signs are present in a considerable number of patients with ASC and necessitate a full spectrum assessment for effective treatment matching (Malviya & Chandra, 2022). This understanding highlights the complex neurological landscape ASC individuals navigate, potentially exacerbating the challenge of differentiating virtual from real environments.

Technological advancements aim to mitigate these challenges by increasing the realism of virtual environments and using embodied cognition techniques to help the brain understand the virtual world as a body-based experience. Enhancing the resolution, frame rate of virtual reality headsets, and employing haptic feedback can make virtual environments more brain-friendly. Such interventions are crucial for neurodivergent individuals as they navigate both real and increasingly prevalent virtual environments, ensuring they can effectively engage with and benefit from the technology around them. Understanding these preferences not only enhances our appreciation of the diversity of aesthetic experiences but also informs the creation of more accessible and enjoyable artistic environments for individuals with different neurotypes. As we continue to explore and understand these preferences, it becomes increasingly important to consider them in the broader context of art education, creation, and appreciation, ensuring that all individuals can fully engage with and enjoy the rich world of art.

3.4 Identifying Gaps in Current Technological Applications

The conditions discussed in this chapter are "invisible" to the public and those that interact with individuals that have them, unless disclosed (Syma, 2019). Under the Americans with Disabilities Act (ADA), a person with a disability is defined as someone with a physical or mental impairment that substantially limits one or more major life activities, however, it should be noted that physical disabilities often receive more accommodations and recognition compared to mental impairments (Emens, 2020; Hutson & Hutson, 2023a). Neurodivergent populations, particularly autistic individuals, face challenges in obtaining appropriate support for physical conditions, impacting their educational and social development (Donaghy et al., 2023). As such, identifying gaps in current technological applications must also recognize the impact of *ableism* and invisible conditions on neurodivergent populations. Ableism, a set of beliefs or practices that devalue people with disabilities, often renders individuals with invisible conditions like neurodivergence less visible and undermines their experiences and opportunities in society (Hidegh et al., 2022). The societal bias affects neurodivergent individuals by influencing their responses to social challenges and limiting the acknowledgment and support they receive often considered to be "faking" their conditions (Boue, 2022). People with "invisible" physical disabilities, including neurodivergent conditions, frequently encounter microaggressions, body policing, and tension in roles due to ableism. Similarly, disabled embodiment is produced and experienced within an ableist context, focusing more on impaired bodies rather than the recognition and celebration of disabled identity (Loja et al., 2013). These experiences contribute to a desire for justice and, often, internalized ableism, which can profoundly affect their well-being and self-perception (Kattari et al., 2018).

To address these challenges, a post-ableist approach in all spheres must be adopted, including technology and art. This involves creating safer spaces and affirming the identities of individuals with disabilities, neurodivergence, and chronic physical and mental health conditions (Shaw et al., 2022). Incorporating neurodivergence into therapeutic practice can help disabled individuals protect their well-being against ableism by identifying with their ingroup and rejecting ableism as illegitimate (Chapman & Botha, 2023). By recognizing and actively working against ableism, the gaps in current technological applications can be closed, ensuring that the needs and preferences of neurodivergent populations are catered to, not only improving accessibility and inclusion but also enriches the diversity and innovation in technology and art, benefiting society as a whole (Stallings, 2022).

Understanding gender, age, and sensory perception is crucial in addressing disparities in technological solutions. Aging significantly influences sensory thresholds, affecting how older individuals interact with technology and perceive sensory stimuli. As age increases, sensory thresholds for various modalities, including tactile, auditory, and visual, tend to rise, leading to changes in perception and interaction dynamics (O'Dowd et al., 2022). The elevation in thresholds means older adults may require stronger stimuli to register sensations, which can affect their use and experience of technology designed with standard sensory levels in mind (Rhodus et al., 2022). For instance, older adults exhibit raised tactile thresholds, necessitating adjustments in devices that rely on touch or haptic feedback (Martiniello et al., 2022). Similarly, aging leads to changes in the perception of thermal stimuli, with modest changes in warming and cooling perception, indicating that sensory experiences can become less acute or differ in nature as one ages (Coull et al., 2022). These changes have profound implications for designing user interfaces, virtual reality experiences, and other technologies that depend on sensory input. The challenge is further compounded by the fact that aging affects sensory modalities differently, with some undergoing more significant modifications than others (Alain et al., 2022). This variability requires a nuanced understanding of how aging impacts sensory perception to design effective technological solutions. Moreover, sensory decline and information integration independently explain cognitive status in healthy older adults, emphasizing the need for comprehensive strategies that address both sensory and cognitive changes with age (Kiss et al., 2023).

Aside from age, gender is also an important factor to consider. First, gender has proven problematic when it comes to diagnoses for certain conditions that limit intervention and addressing sensory processing challenges (Lai et al., 2022). In particular, females receive far less research and diagnoses compared to their male counterparts and this is a critical concern in understanding and addressing the condition (Hutson & Hutson, 2023b). The evidence supports the existence of a female-typical autism presentation, but its characteristics and impact across all genders and ages require further examination (Hull et al., 2021). There is a lack of understanding and recognition of the smaller gender ratio of autism in patients with intellectual disability, with many healthcare professionals feeling less confident in recognizing, screening, and diagnosing autism in female patients (Tromans et al., 2021). Girls and women are nearly invisible in research and often underserved, with the ratio of boys to girls with ASC being 4:1 overall and even higher for autism without intellectual disability (Rynkiewicz & Łucka, 2018). Autistic females often have more developed social skills than males, which may contribute to their under-diagnosis (Gould, 2017). Additionally, autistic females present differently than males, and low recognition affects their diagnosis, assessment reliability, and access to support (Driver & Chester, 2021). The discrepancy leads to those who identify as females often receiving diagnoses later than males, resulting in misunderstood needs and experiences of self-acceptance and understanding (Davidovitch et al., 2023).

Gender differences in technology engagement and educational settings significantly influence how individuals interact with and perceive technology. In the realm of education, girls exhibit more interest in software structure and design compared to boys, who show more interest in computer attitudes (Dong & Zhu, 2023). Female students using mobile technology in group learning modes supported by video material demonstrate higher peer-influenced learning scores and engagement patterns compared to male students (Zhou et al., 2023). The results suggest that gender plays a role in how students engage with and learn from technology. In constructionist classrooms, girls using technologies like Lego/Logo show significant gains in self-efficacy and positive perceptions of self, challenging traditional masculine perceptions of careers in science and technology (Pampaliari et al., 2023). Additionally, females are more likely to use student-centered pedagogical approaches in teaching but often have lower confidence and less experience using computers in teaching, indicating a need for targeted professional development (MacMaoilir & McGillicuddy, 2023). Moreover, identity development of females in computing and technology classrooms is influenced by heterosexual social norms, necessitating a broader understanding of gender and technology relations in education (Trauth, 2023). Despite these insights, girls aged 9–12 confirm male norms and conceptions of technology, but also resist the self-image of not being technical when working together and having ownership of their technology learning (Sultan et al., 2023). The technological gender gap persists, and educators and designers of computer-based instruction should address it to ensure equitable educational opportunities for both male and female students (Delaney & Devereux, 2022).

Age and gender significantly influence sensory perception, which directs interactions with healthcare and technology. Age-related differences in current perception thresholds (CPT) are observed in both genders, indicating that as individuals age, their sensitivity to various sensory inputs changes (Zhang et al., 2021). This has notable implications for how older individuals perceive and interact with technology, as they may require stronger or more distinct stimuli to achieve the same level of perception as younger individuals. Additionally, older individuals experience changes in esophageal biomechanical properties and sensation, affecting their digestive functions and potentially their nutritional intake and comfort (Feng et al., 2023). These changes occur without significant differences between genders, suggesting that aging itself plays a critical role in sensory alteration. Gender-specific differences also exist in sensory perception, with men and women experiencing and processing sensory information differently. For example, males have a higher CPT at certain frequencies, indicating a difference in how sensory information is perceived at the neurological level (Akins et al., 2023).

The salience of gender considerations for technology, particularly in the context of sensory disabilities and aesthetic experiences, stems from the fact that gender differences can significantly limit the effectiveness of technological interventions for females (Mariscal et al., 2019). Women may experience and interact with technology differently due to variations in sensory perception, cognitive processing, or even cultural and societal expectations (Kalender et al., 2019). If technological solutions are designed without considering these differences, they might not effectively address or accommodate the specific needs or preferences of female users, leading to less optimal or even adverse experiences. For instance, if a therapeutic application relies on sensory stimuli that do not account for the typically higher sensory thresholds or different sensory processing patterns in women, it may be less effective or even uncomfortable for them, thereby reducing the overall efficacy of the intervention. Similarly, aesthetic experiences mediated by technology, such as VR or digital art platforms, must consider gender-specific preferences and perceptual differences to ensure engaging and meaningful experiences for all users (MacArthur et al., 2021).

Given these technical and demographic considerations, therapeutic and adaptive applications for sensory disabilities are continually evolving, aiming to enhance the quality of life and functional abilities of individuals with various sensory profiles. Sensory-integrative therapy, for instance, effectively improves balance, muscle tone, and social engagement in children with sensory-integrative conditions through a combination of vestibular stimulation and movement activities (Hafez Afefe Barakat et al., 2023). Similarly, Snoezelen, a multisensory stimulation technique, shows potential in promoting adaptive behaviors in older people with dementia (Carrera-Hidalgo & Zhimnay-Valverde, 2020). Sensory integration therapy has also been used to treat self-injurious behavior in adults with profound mental retardation by employing multiple short sessions spread throughout the day (Dao et al., 2023). Furthermore, motor imagery-based brain–computer interfaces are emerging as a technology to rehabilitate motor deficits, potentially leading to implantable devices and a focus on sensory-motor integration (Padfield et al., 2019).

While sensory integration therapy shows promise, its effectiveness as a primary or adjunctive treatment for children with learning disabilities and other disorders is still under debate due to current research findings. Some studies suggest that the therapy may be ineffective or its application has methodological flaws, highlighting the need for improved research methods (Schoen et al., 2019). Sensory neuroprostheses are another avenue of research, showing promise in restoring lost function for patients with severe sensory disabilities (Fig. 3.11). These require efficient delivery of sensory information, effective rehabilitation, and lifelong follow-up (Charkhkar et al., 2020). Additionally, using specialist sensory equipment in adult hospices can enhance health and well-being for patients and visitors, with positive feedback from staff and patients. Alternatively, occupational therapy interventions aim to enhance daily living skills, promote independence, and enhance the quality of life for individuals with disabilities and their families (Reitz et al., 2020). These interventions may focus on sensory skills, physical, mental, and emotional disabilities, and often involve a multidisciplinary approach for effective outcomes (Howard et al., 2020).

Beyond general interventions, those specific to art therapy and neuroarts utilize sensory stimulation to enhance brain functions related to visual feature recognition, emotional states, and memory processing, providing therapeutic and adaptive applications for individuals with sensory disabilities (Canbeyli, 2022). These interventions, rooted in the principles of neuroarts, leverage the brain's response to artistic and sensory stimuli to facilitate healing, learning, and growth. Multisensory stimulation methods, such as motor imagery and music therapy, are particularly effective in stroke rehabilitation, enhancing brain processing and the ability to recognize sensory stimuli. These techniques help retrain the brain and improve motor functions by engaging multiple senses simultaneously (Haire, 2021). Active sensing, which involves motor and

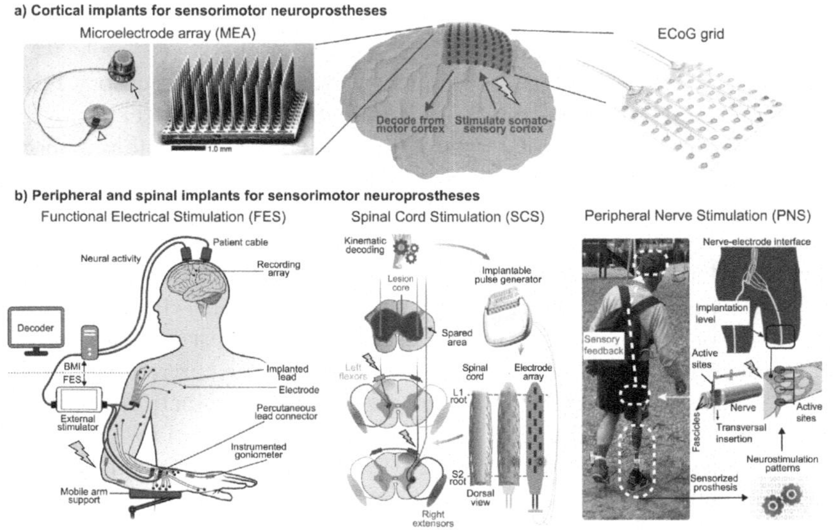

Fig. 3.11 Neuroprosthetic technologies for sensorimotor disorders (*Source* From Wikimedia Commons, licensed under CC-BY 4.0)

attentional sampling routines, indicates that sensory processing is an active and rhythmic endeavor. This understanding is crucial for designing interventions that manipulate attentional rhythms for perceptual selection, potentially benefiting individuals with sensory processing difficulties (Leszczynski & Schroeder, 2019).

In the context of neurodevelopmental disorders, auditory-based targeted cognitive training has shown improvements in working memory and information processing speed in autistic adolescents and young adults. This suggests that auditory sensitivities and the brain's plasticity can be harnessed for therapeutic interventions (Vinogradov, 2023). Additionally, VR sensory rooms are being explored as tools for supporting adults with neurodevelopmental disabilities, although optimal outcomes require further research and support staff training (Mills et al., 2023). Sensory processing interventions should be incorporated into daily routines of individuals and environments to yield generalized outcomes. The Sensory Processing Scale for Monkeys (SPS-M) effectively assesses sensory processing in adult rhesus monkeys exposed to prenatal alcohol, stress, or postnatal lead, aiding in occupational therapy treatment planning

(Schneider et al., 2007). Understanding sensory processing in autism is also enhanced by neurophysiological profiles, which may serve as valuable biomarkers for diagnosis and monitoring therapeutic interventions (Klin, 2018). Overall, therapeutic and adaptive applications in neuroarts and sensory processing interventions offer a range of techniques and strategies to support individuals with sensory disabilities. By understanding the underlying mechanisms of sensory processing and leveraging the therapeutic potential of art and multisensory stimulation, these interventions can significantly improve the lives of individuals with sensory impairments, contributing to a more inclusive and supportive environment.

Aside from clinical environments, museums have also been implementing various sensory adaptations to create inclusive environments for all visitors, including those with sensory processing disorders or neurodiverse conditions like autism. Sensory bags and sensory-friendly days are examples of such adaptations aimed at making the museum experience more accessible and enjoyable. Sensory bags typically include items like noise-canceling headphones, fidget tools, and visual cue cards. These tools help visitors who might be overwhelmed by the sensory environment of a museum, allowing them to regulate their sensory input and engage with the exhibits in a more comfortable way (Chatterjee & Vartanian, 2014). Sensory-friendly days or hours are another adaptation, where museums may adjust lighting and sound levels, provide quiet spaces, and make staff aware of the needs of visitors with sensory sensitivities (Schwartzman & Knowles, 2022).

However, while these adaptations are valuable, they also have limitations. For instance, sensory bags might not cater to the specific needs or preferences of all visitors with sensory sensitivities, and the items included may not be effective for everyone. Sensory-friendly days may only occur occasionally, limiting access for those who would benefit from them. Additionally, these adaptations might not address all types of sensory sensitivities or the full range of neurodivergent conditions, underscoring the need for a broader, more nuanced approach. Understanding the effectiveness of these interventions is crucial. Research suggests that while these adaptations are well-intentioned, continuous feedback and adaptation are necessary to meet the diverse needs of all visitors effectively (Hutson & Hutson, 2023c). Museums must engage with communities, experts, and visitors themselves to understand better and address the barriers within the museum environment. Therefore, while sensory bags and sensory-friendly days are positive steps toward inclusivity, ongoing research, dialogue, and innovation are needed to ensure that museums can provide a welcoming and engaging experience for all visitors, regardless of sensory needs or neurodivergent conditions. This involves not only providing sensory adaptations but also cultivating an environment of understanding and acceptance which permeates all aspects of the museum experience.

In considering the diverse sensory experiences and aesthetic preferences associated with various neurodivergent conditions and disabilities, the profound influence these have on individual interaction with art and technology is profound. From the complexities of prosopagnosia and dyscalculia to the unique challenges and possibilities presented by aphantasia and sensory adaptations in museums, the chapter underscores the need for more inclusive, sensitive, and innovative approaches in both therapeutic interventions and technological applications. Despite progress in understanding and accommodating these conditions, significant gaps remain, particularly in gender and age-related disparities, as well as in the equitable recognition and support for neurodivergent individuals. Moving forward, it is clear a deeper understanding of the diagnostic criteria for various synesthetes and neurotypes is necessary to continue advancing our knowledge and improving support systems. Chapter 4 aims to build on this foundation, investigating diagnostic criteria and further exploring the intersections of neurodiversity, sensory perception, and art, ensuring that the benefits of technological and therapeutic advancements reach all individuals, regardless of their sensory and cognitive profiles.

References

á Brekku, L. (2019). *The diversity of color through color theory* (Doctoral dissertation).

Agostini, F., Zoccolotti, P., & Casagrande, M. (2022). Domain-general cognitive skills in children with mathematical difficulties and dyscalculia: A systematic review of the literature. *Brain Sciences, 12*(2), 239.

Akins, J. D., Okada, Y., Hendrix, J. M., Vongpatanasin, W., & Fu, Q. (2023). Greater resting muscle sympathetic nerve activity reduces cold pressor autonomic reactivity in older women but not older men. *American Journal of Physiology-Regulatory, Integrative and Comparative Physiology, 324*(5), R656–R665.

Alain, C., Chow, R., Lu, J., Rabi, R., Sharma, V. V., Shen, D., ... & Freedman, M. (2022). Aging enhances neural activity in auditory, visual, and somatosensory cortices: The common cause revisited. *Journal of Neuroscience, 42*(2), 264–275.

Alashram, A. R., & Annino, G. (2023). A novel neurorehabilitation approach for neural plasticity overstimulation and reorganization in patients with neurological disorders. *Physikalische Medizin, Rehabilitationsmedizin, Kurortmedizin.*

Albert, B., Zanni-Merk, C., Beuvron, F., Maire, J., Pillet, M., Charrier, J., & Knecht, C. (2016). *A smart system for haptic quality control—Introducing an ontological representation of sensory perception knowledge* (pp. 21–30). https://doi.org/10.5220/0006048300210030

Alcaraz Martinez, R., Turró, M. R., & Granollers Saltiveri, T. (2022). Methodology for heuristic evaluation of the accessibility of statistical charts for people with low vision and color vision deficiency. *Universal Access in the Information Society, 21*(4), 863–894.

Alink, A., & Charest, I. (2020). Clinically relevant autistic traits predict greater reliance on detail for image recognition. *Scientific Reports, 10*(1), 14239.

Allen, B. (2023). *Living in time: The philosophy of Henri Bergson.* Oxford University Press.

Alper, M. (2018). Inclusive sensory ethnography: Studying new media and neurodiversity in everyday life. *New Media & Society, 20*, 3560–3579. https://doi.org/10.1177/1461444818755394

Arcangeli, M. (2023). Aphantasia demystified. *Synthese, 201*(2), 31.

Arizpe, J. M., Saad, E., Douglas, A. O., Germine, L., Wilmer, J. B., & DeGutis, J. M. (2019). Self-reported face recognition is highly valid, but alone is not highly discriminative of prosopagnosia-level performance on objective assessments. *Behavior Research Methods, 51*, 1102–1116.

Arnaud, S., & Pendoley, K. (2023). Personal intentionalism and the understanding of emotion experience. *Journal of Consciousness Studies, 30*(7–8), 61–87.

Arshamian, A., Gerkin, R. C., Kruspe, N., Wnuk, E., Floyd, S., O'Meara, C., ... & Majid, A. (2022). The perception of odor pleasantness is shared across cultures. *Current Biology, 32*(9), 2061–2066.

Aru, J., Larkum, M. E., & Shine, J. M. (2023). The feasibility of artificial consciousness through the lens of neuroscience. *Trends in Neurosciences.*

Aslan, I. H., Grant, J. E., & Chamberlain, S. R. (2023). Cognition in adults with borderline personality disorder. *CNS Spectrums*, 1–6.

Avidan, G., & Behrmann, M. (2021). Spatial integration in normal face processing and its breakdown in congenital prosopagnosia. *Annual Review of Vision Science, 7*, 301–321.

Barbur, J. L., Petzold, A., & Plant, G. T. (2022). Colour-related, binocular interactions in patients with dyschromatopsia. *Acta Ophthalmologica*, 100.

Bartolomeo, P., & Miceli, G. (2023). *Object color agnosia*. Cortex.

Barton, J. J., Davies-Thompson, J., & Corrow, S. L. (2021). Prosopagnosia and disorders of face processing. In *Handbook of clinical neurology* (Vol. 178, pp. 175–193). Elsevier.

Bhattacharya, T., Dasgupta, S., Saha, T., & Sen, J. (2020). Altered perception of culture: Based on features of pedestrian experience and aesthetic regeneration of built environment (pp. 447–471). https://doi.org/10.1007/978-3-030-25879-5_19

Blom, J. D., Ter Meulen, B. C., & Dool, J. (2021). A century of prosopometamorphopsia studies. *Cortex, 139*, 298–308.

Boesveldt, S., Postma, E. M., Boak, D., Welge-Luessen, A., Schöpf, V., Mainland, J. D., ... & Duffy, V. B. (2017). Anosmia—A clinical review. *Chemical Senses, 42*(7), 513-523.

Bouchard, S., & Rizzo, A. S. (2019). Applications of virtual reality in clinical psychology and clinical cognitive neuroscience–An introduction. *Virtual reality for psychological and neurocognitive interventions*, 1–13.

Boue, S. (2022). Not all surfaces catch the light at the same time. *Neurodivergence and Architecture*, 223.

Bracey, A. (2020). Painting Aphantasia: Making the non-visible, visible. *PhotographyDigitalPainting: Expanding medium interconnectivity in contemporary visual art practices*, 99.

Burke, P. (2020). *The polymath: A cultural history from Leonardo da Vinci to Susan Sontag*. Yale University Press.

Calzavarini, F. (2023). Rethinking modality-specificity in the cognitive neuroscience of concrete word meaning: A position paper. *Language, Cognition and Neuroscience*, 1–23.

Canbeyli, R. (2022). Sensory stimulation via the visual, auditory, olfactory and gustatory systems can modulate mood and depression. *European Journal of Neuroscience, 55*(1), 244–263.

Carrera-Hidalgo, P., & Zhimnay-Valverde, E. O. (2020, November). An expert system to design intervention strategies for children with disabilities in multi-sensory stimulation environments. In *Information and communication technologies: 8th conference, TICEC 2020, Guayaquil, Ecuador, November 25–27, 2020, Proceedings* (Vol. 1307, p. 51). Springer Nature.

Casciano, F., Zauli, E., Rimondi, E., Mura, M., Previati, M., Busin, M., & Zauli, G. (2022). The role of the mTOR pathway in diabetic retinopathy. *Frontiers in Medicine, 9*, 973856.

Cascio, C. J., Moana-Filho, E. J., Guest, S., Nebel, M. B., Weisner, J., Baranek, G. T., & Essick, G. K. (2012). Perceptual and neural response to affective tactile texture stimulation in adults with autism spectrum disorders. *Autism Research, 5*(4), 231–244.

Cárdenas-Belaunzarán, J., & Cerrillo-Avila, K. A. (2023). Visual agnosia mimicking memory impairment: A case report of posterior cortical atrophy. *Neuro-Ophthalmology*, 1–7.

Cattaneo, M. (1994). Addressing culture and values in the training of art therapists. *Art Therapy, 11*, 184–186. https://doi.org/10.1080/07421656.1994.10759081

Chang, C. H., Liu, C. Y., Chen, S. J., & Tsai, H. C. (2018). Efficacy of light therapy on nonseasonal depression among elderly adults: A systematic review and meta-analysis. *Neuropsychiatric Disease and Treatment*, 3091–3102.

Chapman, R., & Botha, M. (2023). Neurodivergence-informed therapy. *Developmental Medicine & Child Neurology, 65*(3), 310–317.

Charkhkar, H., Christie, B. P., & Triolo, R. J. (2020). Sensory neuroprosthesis improves postural stability during Sensory Organization Test in lower-limb amputees. *Scientific Reports, 10*(1), 6984.

Chatterjee, A. (2006). The neuropsychology of visual art: Conferring capacity. *International Review of Neurobiology, 74*, 39–49.

Chatterjee, A., & Vartanian, O. (2014). Neuroaesthetics. *Trends in Cognitive Sciences, 18*(7), 370–375.

Cheung, V. K., Harrison, P. M., Koelsch, S., Pearce, M. T., Friederici, A. D., & Meyer, L. (2024). Cognitive and sensory expectations independently shape musical expectancy and pleasure. *Philosophical Transactions of the Royal Society B, 379*(1895), 20220420.

Chen, W., Song, J., Wang, Y., Wu, C., Ma, S., Wang, D., ... & Li, H. (2023). Inattentional blindness to unexpected hazard in augmented reality head-up display assisted driving: The impact of the relative position between stimulus and augmented graph. *Traffic Injury Prevention, 24*(4), 344–351.

Cheron, G., Ristori, D., Petieau, M., Simar, C., Zarka, D., & Cebolla, A. M. (2022). Effects of pulsed-wave chromotherapy and guided relaxation on the theta-alpha oscillation during arrest reaction. *Frontiers in Psychology, 13*, 792872.

Classen, C. (1990). Sweet colors, fragrant songs: Sensory models of the Andes and the Amazon. *American Ethnologist, 17*, 722–735. https://doi.org/10.1525/AE.1990.17.4.02A00070

Classen, C. (1993). *Worlds of Sense: Exploring the senses in history and across cultures.*https://doi.org/10.2307/3341985

Conway, B. R. (2022). Vision and art. In *Oxford research encyclopedia of neuroscience*. Oxford University Press.

Coull, N. A., Hodder, S. G., & Havenith, G. (2022). Age comparison of changes in local warm and cold sensitivity due to whole body cooling. *Journal of Thermal Biology, 104*, 103174.

Courte, J., Le, N. A., Pan, T., Bousset, L., Melki, R., Villard, C., & Peyrin, J. M. (2023). Synapses do not facilitate prion-like transfer of alpha-synuclein:

A quantitative study in reconstructed unidirectional neural networks. *Cellular and Molecular Life Sciences, 80*(10), 284.
D'Aloisio-Montilla, N. (2017). Imagery and overflow: We see more than we report. *Philosophical Psychology, 30*(5), 545–570.
Dao, V., Guetta, M., Giannitelli, M., Doulou, F., Leullier, M., Ghattassi, Z., ... & Cohen, D. (2023). Severe self-injurious behaviors in an autistic child with sensory seeking, depressive disorder and anxiety disorder: A focus on the therapeutic interventions. *Neuropsychiatrie de l'Enfance et de l'Adolescence.*
Davidovitch, M., Gazit, S., Patalon, T., Leitner, Y., & Rotem, R. S. (2023). Late diagnosis of autism spectrum disorder—Journey, parents' concerns, and sex influences. *Autism Research, 16*(2), 294–301.
Dawes, A. J., Keogh, R., Andrillon, T., & Pearson, J. (2020). A cognitive profile of multi-sensory imagery, memory and dreaming in aphantasia. *Scientific Reports, 10*(1), 10022.
Decarli, G., Sella, F., Lanfranchi, S., Gerotto, G., Gerola, S., Cossu, G., & Zorzi, M. (2023). Severe developmental dyscalculia is characterized by core deficits in both symbolic and nonsymbolic number sense. *Psychological Science, 34*(1), 8–21.
Dedrick, D. (2022). Colour classification in natural languages. *Knowledge Organization, 48*(7–8), 563–579.
de Gelder, B., Huis in 't Veldt, E., Zhan, M., & Van den Stock, J. (2022). Acquired prosopagnosia with structurally intact and functional fusiform face area and with face identity-specific configuration processing deficits. *Cerebral Cortex, 32*(21), 4671–4683.
Deh, T. H. (2019). Thinking in pictures: A way forward for dyslexia, dysgraphia and dyscalculia in Ghana. *European Journal of Public Health Studies, 1*(2).
Dehghani, H. (2019, December). The effectiveness of a mobile application "Kalcal" on the learning of mathematics in students with dyscalculia. In *2019 International Serious Games Symposium (ISGS)* (pp. 1–6). IEEE.
Delaney, J. M., & Devereux, P. J. (2022). Gender differences in STEM persistence after graduation. *Economica, 89*(356), 862–883.
DeNora, T. (2014). Making sense of reality: Culture and perception in everyday life. *Making Sense of Reality*, 1–200.
Derke, F., Filipović-Grčić, L., Raguž, M., Lasić, S., Orešković, D., & Demarin, V. (2023). Neuroaesthetics: How we like what we like. *Mind, brain and education* (pp. 1–12). Springer International Publishing.
De Renzi, E. (2020). Agnosia. In *Handbook of clinical and experimental neuropsychology* (pp. 371–408). Psychology Press.
Dong, Y., & Zhu, S. (2023). Gender differences in creative design education: Analysis of individual creativity and artefact perception in the first-year design studio. *International Journal of Technology and Design Education, 33*(1), 165–189.

Donaghy, B., Moore, D., & Green, J. (2023). Co-occurring physical health challenges in neurodivergent children and young people: A topical review and recommendation. *Child Care in Practice, 29*(1), 3–21.

Dorriy, Z., & Shool, S. (2019). Psychological Impact of Colors on 7–11 years Hyperactive Children. *International Journal of Applied Arts Studies (IJAPAS), 4*(2), 25–32.

Driver, B., & Chester, V. (2021). The Recognition and diagnosis of autism in women and girls: A literature review. *Advances in Autism, 7*(3), 194–207.

Eagleman, D. M., & Perrotta, M. V. (2023). The future of sensory substitution, addition, and expansion via haptic devices. *Frontiers in Human Neuroscience, 16*, 1055546.

Edkins, J. (2013). Dismantling the face: Landscape for another politics? *Environment and Planning d: Society and Space, 31*(3), 538–553.

Elkoshi, R. (2019). When sounds, colors, and shapes meet: Investigating children's audiovisual art in response to classical music. *International Journal of Music Education, 37*(4), 576–592.

Elmarakbi, N., Pearson, A., & McIntyre, J. (2023). Creative design thinking approach to support the complex learning environment of the classroom for autistic children and their teachers. In *DS 124: Visual proceedings of the 25th International conference on engineering and product design education (E&PDE 2023).*

Emens, E. F. (2020). *Disability admin: The invisible costs of being disabled*, 2329.

Enfield, N. J. (2022). Utilitarian versus Intellectualist Explanations of Lexical Content: A False Dichotomy. In *The art of language* (pp. 423–438). Brill.

Eslami, M. J., Etemadifar, M., Khalili, M. R., Faraidan-Esfahani, M., Jahanbani-Ardakani, H., Ghorbani, S., & Abtahi, S. H. (2022). Do optic neuritis patients report their visual impairment more frequently upon awakening? *International Journal of Neuroscience, 132*(12), 1261–1263.

Fan, J. E., Bainbridge, W. A., Chamberlain, R., & Wammes, J. D. (2023). Drawing as a versatile cognitive tool. *Nature Reviews Psychology, 2*(9), 556–568.

Farah, M. J., Levinson, K. L., & Klein, K. L. (1995). Face perception and within-category discrimination in prosopagnosia. *Neuropsychologia, 33*(6), 661–674.

Fiene, L., Ireland, M. J., & Brownlow, C. (2018). The Interoception Sensory Questionnaire (ISQ): A scale to measure interoceptive challenges in adults. *Journal of Autism and Developmental Disorders, 48*(10), 3354–3366.

Feng, H. Y., Zhang, P. P., & Wang, X. W. (2023). Presbyphagia: Dysphagia in the elderly. *World Journal of Clinical Cases, 11*(11), 2363.

Finlay, K. (2020). *A world without smell: A visual exploration of an olfactory issue* (Doctoral dissertation, Auckland University of Technology).

Fivush, R., & Grysman, A. (2023). Accuracy and reconstruction in autobiographical memory:(Re) consolidating neuroscience and sociocultural developmental approaches. *Wiley Interdisciplinary Reviews: Cognitive Science, 14*(3), e1620.

Fox-Muraton, M. (2021). Aphantasia and the language of imagination: A wittgensteinian exploration. *Analiza i Egzystencja: Czasopismo Filozoficzne, 55*, 5–24.

Freedburg, D. (2014). Feelings on faces. *Rethinking Emotion*, 289–323.

Fretwell, E., & Hsu, H. L. (2023). Redistributions of the sensible: An introduction to "Senses with/out subjects." *American Literature, 95*(3), 447–457.

Furman, M., Fleitas-Rumak, P., Lopez-Segura, P., Furman, M., Tafet, G., de Erausquin, G. A., & Ortiz, T. (2022). Cortical activity involved in perception and imagery of visual stimuli in a subject with aphantasia An EEG Case Report. *Neurocase, 28*(4), 344–355.

Gage, J., & Grovier, K. (2023). *Colour in art*. Thames & Hudson.

Galison, P. (1997). *Image and logic: A material culture of microphysics*. University of Chicago Press.

Gallo, P. G., Panza, M., Viviani, F., & Lantieri, P. B. (1998). Congenital dyschromatopsia and school achievement. *Perceptual and Motor Skills, 86*(2), 563–569.

George, J., Greene, B., & Blackwell, M. (2005). Three Voices on multiculturalism in the art therapy classroom. *Art Therapy, 22*, 132–138. https://doi.org/10.1080/07421656.2005.10129492

Gibson, J. (1963). The useful dimensions of sensitivity. *American Psychologist, 18*, 1–15. https://doi.org/10.1037/H0046033

Gillespie, K. (2022). Facing geography: A new research agenda. *Progress in Human Geography, 46*(1), 198–214.

Göbel, S. M., Terry, R., Klein, E., Hymers, M., & Kaufmann, L. (2022). Impaired arithmetic fact retrieval in an adult with developmental dyscalculia: Evidence from behavioral and functional brain imaging data. *Brain Sciences, 12*(6), 735.

Gopnik, B. (2017, July 13) Jeff Koons: One of today's 10 most important artists. *Daily Beast*. https://www.thedailybeast.com/jeff-koons-one-of-todays-10-most-important-artists

Gulnoza, R., Mushtariy, R., & Aziza, Q. (2023). The intricacies of perception: Exploring physiological foundations, properties, and nerve conditions. *Journal of Academic Research and Trends in Educational Sciences, 2*(2), 278–282.

Gopinath, N. (2023). Artificial intelligence and neuroscience: An update on fascinating relationships. *Process Biochemistry, 125*, 113–120.

Gould, J. (2017). Towards understanding the under-recognition of girls and women on the autism spectrum. *Autism, 21*(6), 703–705.

Griffa, A., Mach, M., Dedelley, J., Gutierrez-Barragan, D., Gozzi, A., Allali, G., ... & Amico, E. (2023). Evidence for increased parallel information transmission in human brain networks compared to macaques and male mice. *Nature Communications, 14*(1), 8216.

Habbak, A. L. Z., & Khodeir, L. (2023). Multi-sensory interactive interior design for enhancing skills in children with autism. *Ain Shams Engineering Journal, 14*(8), 102039.

Hafez Afefe Barakat, A., Ata Abd El-Salihen, F., Mohamed Khalifa, A., & Mohammed Ibrahim, F. (2023). Sensory integrative intervention: Outcomes of motor, cognitive and social skills among children with attention deficit hyperactivity disorders. *Egyptian Journal of Health Care, 14*(3), 726–738.

Haire, C. M. (2021). *Therapeutic Instrumental music training and sensory-enhanced motor imagery in chronic post-stroke upper-extremity rehabilitation: A randomized controlled trial.* University of Toronto, Canada.

Hamilton-Fletcher, G., Alvarez, J., Obrist, M., & Ward, J. (2022). SoundSight: a mobile sensory substitution device that sonifies colour, distance, and temperature. *Journal on Multimodal User Interfaces*, 1–17.

Hart, E., & Hay, L. (2022). Do you see what I see? Exploring vividness of visual mental imagery in product design ideation. *Proceedings of the Design Society, 2*, 881–890.

Hassanpoor, N., Ebrahimiadib, N., & Hajizadeh, F. (2023). Hereditary disease. *Atlas of ocular optical coherence tomography* (pp. 365–415). Springer International Publishing.

Hauck, P., von Castell, C., & Hecht, H. (2022). Crossmodal correspondence between music and ambient color is mediated by emotion. *Multisensory Research, 35*(5), 407–446.

He, H., Li, J., Xiao, Q., Jiang, S., Yang, Y., & Zhi, S. (2019). Language and color perception: Evidence from Mongolian and Chinese speakers. *Frontiers in Psychology, 10*, 551.

Hidegh, A. L., Svastics, C., Győri, Z., & Csillag, S. (2022). The lived experience of freedom among entrepreneurs with disabilities. *International Journal of Entrepreneurial Behavior & Research, 28*(9), 357–375.

Hite, R., Childers, G., Jones, G., Corin, E., & Pereyra, M. (2021). Describing the experiences of students with ADHD learning science content with emerging technologies. *Journal of Science Education for Students with Disabilities, 24*(1), 12.

Hopkins, C., Surda, P., & Kumar, N. (2020). Presentation of new onset anosmia during the COVID-19 pandemic. *Rhinology, 58*(3), 295–298.

Howard, A. R. H., Lynch, A. K., Call, C. D., & Cross, D. R. (2020). Sensory processing in children with a history of maltreatment: An occupational therapy perspective. *Vulnerable Children and Youth Studies, 15*(1), 60–67.

Huang, Z., & Huang, C. (2021, December). Ways of Viewing in the era of copying image: Study on chuck close's pixel portrait paintings. In *The 5th international conference on art studies: Research, experience, education (ICASSEE 2021)* (Vol. 1, pp. 195–201). Amsterdam University Press.

Huisman, J. L., van Hout, R., & Majid, A. (2022). Stability and change in the colour lexicon of the Japonic languages. *Studies in Language, 46*(2), 323–351.

Hull, L., Petrides, K. V., & Mandy, W. (2021). Cognitive predictors of self-reported camouflaging in autistic adolescents. *Autism Research, 14*(3), 523–532.

Hussain, K., & Soares, N. (2022). Dyscalculia. *International Journal of Child Health & Human Development, 15*(3).

Hutson, J., & Hutson, P. (2023a). Digital inclusion for people with autism spectrum disorders: Review of the current legal models and doctrinal concepts. *Journal of Digital Technologies and Law, 1*(4), 851–879.

Hutson, P., & Hutson, J. (2023b). Autism in Females: Understanding the overlooked diagnoses, unique challenges, and recommendations. *Journal of Clinical & Medical Images, 3*(3).

Hutson, J., & Hutson, P. (2023c). *Museums and the Metaverse: Emerging technologies to promote inclusivity and engagement.* Intech Open.

Hutson, J., & McGinley, C. (2023). Neuroaffirmative approaches to extended reality: Empowering individuals with autism spectrum condition through immersive learning environments. *International Journal of Technology in Education and Science, 7*(3).

Iigaya, K., Yi, S., Wahle, I. A., Tanwisuth, K., & O'Doherty, J. P. (2021). Aesthetic preference for art can be predicted from a mixture of low-and high-level visual features. *Nature Human Behaviour, 5*(6), 743–755.

Iosifyan, M., Sidoroff-Dorso, A., & Wolfe, J. (2022). Cross-modal associations between paintings and sounds: Effects of embodiment. *Perception, 51*(12), 871–888.

Ipser, A., Ward, J., & Simner, J. (2020). The Multisense Test of lexical–gustatory synaesthesia: An automated online diagnostic. *Behavior Research Methods, 52*, 544–560.

Jackson, M. B. (2009). Varieties of Intentionalism. In *Intentional Horizons* (pp. 38–74). Brill mentis.

Jeong, S., & Lee, J. (2021). Effects of cultural background on consumer perception and acceptability of foods and drinks: A review of latest cross-cultural studies. *Current Opinion in Food Science, 42*, 248–256. https://doi.org/10.1016/J.COFS.2021.07.004

Jonauskaite, D., Epicoco, D., Al-rasheed, A. S., Aruta, J. J. B. R., Bogushevskaya, V., Brederoo, S. G., ... & Mohr, C. (2023). A comparative analysis of colour–emotion associations in 16–88-year-old adults from 31 countries. *British Journal of Psychology.*

Kalender, Z. Y., Marshman, E., Schunn, C. D., Nokes-Malach, T. J., & Singh, C. (2019). Why female science, technology, engineering, and mathematics majors do not identify with physics: They do not think others see them that way. *Physical Review Physics Education Research, 15*(2), 020148.

Kapitan, L. (2014). Introduction to the neurobiology of art therapy: Evidence based, complex, and influential. *Art Therapy, 31*, 50–51. https://doi.org/10.1080/07421656.2014.911027

Kapitan, L. (2015). Social action in practice: Shifting the ethnocentric lens in cross-cultural art therapy encounters. *Art Therapy, 32*(3), 104–111.

Karim, A. R., Prativa, S., & Likova, L. T. (2022). Perception and appreciation of tactile objects: The role of visual experience and texture parameters. *Journal of perceptual imaging*, 5.

Kattari, S. K., Olzman, M., & Hanna, M. D. (2018). "You look fine!" Ableist experiences by people with invisible disabilities. *Affilia, 33*(4), 477–492.

Keane, G. (2021). Character animation. *Animation & Cartoons*, 50.

Kenter, J. O., & O'Connor, S. (2022). The life framework of values and living as nature; towards a full recognition of holistic and relational ontologies. *Sustainability Science, 17*(6), 2529–2542.

Keogh, R., Pearson, J., & Zeman, A. (2021). Aphantasia: The science of visual imagery extremes. In *Handbook of clinical neurology* (Vol. 178, pp. 277–296). Elsevier.

Kiss, J., Edwards, G., Bouserhal, R., Champagne, E., Belleguic, T., Psyché, V., ... & Charbonneau, A. (2023). Harnessing artificial intelligence for early and evolution of Alzheimer's disease detections and enhancing senior mental health through innovative art-singing therapies: A multidisciplinary approach. *Journal of Community Medicine and Health Solutions*, 4.

Klein, E., & Knops, A. (2023). The two-network framework of number processing: A step towards a better understanding of the neural origins of developmental dyscalculia. *Journal of Neural Transmission, 130*(3), 253–268.

Klin, A. (2018). Biomarkers in autism spectrum disorder: Challenges, advances, and the need for biomarkers of relevance to public health. *Focus, 16*(2), 135–142.

Klopfenstein, T., Kadiane-Oussou, N. J., Toko, L., Royer, P. Y., Lepiller, Q., Gendrin, V., & Zayet, S. (2020). Features of anosmia in COVID-19. *Medecine Et Maladies Infectieuses, 50*(5), 436–439.

Koons, J. (2014). *About Koons*. Thames & Hudson.

Krempel, R. (2018). Color terms and perception. *Filosofia Analítica*, 64.

Kukahiko, K., Fernandez, P., Sang, D. K., Yim, K., Iwane, A., Kalama-Macomber, K., ... & Fleming-Nazara, T. (2020). Pūpūkahi i Holomua: A story of Hawaiian education and a theory of change. *Encounters in Theory and History of Education, 21*, 175–212.

Ku, B. D., & Kim, G. U. (2022). A patient with persistent prosopagnosia after right posterior cerebral artery territorial infarction. *Neurology Asia, 27*(3), 763–766.

Lai, M. C., Lin, H. Y., & Ameis, S. H. (2022). Towards equitable diagnoses for autism and attention-deficit/hyperactivity disorder across sexes and genders. *Current Opinion in Psychiatry, 35*(2), 90–100.

Laing, D. G., Epps, A., & Jinks, A. L. (2021). Chemosensory loss during a traumatic brain injury suggests a central pathway for the rehabilitation of anosmia. *Chemical Senses, 46*, bjab016.

Langdon, C., Laxe, S., Lehrer, E., Berenguer, J., Alobid, I., Quintó, L., ... & Mullol, J. (2021). Loss of smell in patients with traumatic brain injury is associated with neuropsychiatric behavioral alterations. *Brain injury, 35*(11), 1418–1424.

Lemogne, C., Smadja, J., Zerdazi, E. H., Soudry, Y., Robin, M., Berthoz, S., ... & Bonfils, P. (2015). Congenital anosmia and emotion recognition: a case-control study. *Neuropsychologia, 72*, 52–58.

Leška, M. (2023). *Photorealism and hyperrealism in American arts.*

Leszczynski, M., & Schroeder, C. E. (2019). The role of neuronal oscillations in visual active sensing. *Frontiers in Integrative Neuroscience, 13*, 32.

Li, S., Ma, H., Yang, F., & Ding, X. (2023). CGMP signaling in photoreceptor degeneration. *International Journal of Molecular Sciences, 24*(13), 11200.

Li, J. Y., Xu, Y., Wang, D. G., & Wang, J. H. (2023). The interconnection and function of associative memory neurons are upregulated for memory strengthening. *Frontiers in Neural Circuits, 17*, 1189907.

Lifshitz, M., & Luhrmann, T. (2020). Culture and the plasticity of perception. *Behavioral and Brain Sciences, 43*. https://doi.org/10.1017/S0140525X19002887

Liu, X., Li, X., Song, Y., & Liu, J. (2021). Separate and shared neural basis of face memory and face perception in developmental prosopagnosia. *Frontiers in Behavioral Neuroscience, 15*, 668174.

Loh, V. (2023). *Impossible art: Synesthesia, sensory mimesis, and the emergence of cross-modal works of modern art and literature* (Doctoral dissertation, Temple University).

Loja, E., Costa, M. E., Hughes, B., & Menezes, I. (2013). Disability, embodiment and ableism: Stories of resistance. *Disability & Society, 28*(2), 190–203.

MacArthur, C., Grinberg, A., Harley, D., & Hancock, M. (2021, May). You're making me sick: A systematic review of how virtual reality research considers gender & cybersickness. In *Proceedings of the 2021 CHI conference on human factors in computing systems* (pp. 1–15).

MacKisack, M., Zeman, A., Winlove, C., Onians, J., Macpherson, F., & Aldworth, S. (2020). *Extended Imagining: The case of the aphantasic artist.*

MacMaoilir, E., & McGillicuddy, D. (2023). "I'm actually a female empowerer": Student perspectives on a critical pedagogical approach to re/engage at-risk females in school. *Education and Urban Society, 55*(9), 1047–1069.

Maimon, A., Yizhar, O., Buchs, G., Heimler, B., & Amedi, A. (2022). A case study in phenomenology of visual experience with retinal prosthesis versus visual-to-auditory sensory substitution. *Neuropsychologia, 173*, 108305.

Malik-Moraleda, S., Ayyash, D., Gallée, J., Affourtit, J., Hoffmann, M., Mineroff, Z., ... & Fedorenko, E. (2022). An investigation across 45 languages and 12 language families reveals a universal language network. *Nature Neuroscience, 25*(8), 1014–1019.

Malviya, M., & Chandra, J. (2022). A systematic review on prognosis of autism using machine learning techniques. *ECS Transactions, 107*(1), 11623.

Manferdini, E., & Benyamin, J. (Eds.). (2023). *Full Spectrum: Colour in Contemporary Architecture*. Routledge.

Mansfield, J. (2019). *Murrinhpatha morphology and phonology* (Vol. 653). Walter de Gruyter GmbH & Co KG.

Mariscal, J., Mayne, G., Aneja, U., & Sorgner, A. (2019). Bridging the gender digital gap. *Economics, 13*(1), 20190009.

Martin, R., & Wilkins, J. (2022). Creating visually appropriate classroom environments for students with autism spectrum disorder. *Intervention in School and Clinic, 57*(3), 176–181.

Martiniello, N., Barlow, M., & Wittich, W. (2022). Exploring correlates of braille reading performance in working-age and older adults with visual impairments. *Scientific Studies of Reading, 26*(4), 267–286.

Mendelovici, A. (2023). Truth and content in sensory experience. *Oxford Studies in Philosophy of Mind, 3*, 318.

Metzl, E. S. (2022). Art is fun, art is serious business, and everything in between: Learning from art therapy research and practice with children and teens. *Children, 9*(9), 1320.

Miceli, G., & Caccia, A. (2022). Cortical disorders of speech processing: Pure word deafness and auditory agnosia. In *Handbook of Clinical Neurology* (Vol. 187, pp. 69–87). Elsevier.

Miceli, G., & Caccia, A. (2023). The auditory agnosias: A short review of neurofunctional evidence. *Current Neurology and Neuroscience Reports, 23*(11), 671–679.

Michels, L., Buechler, R., & Kucian, K. (2022). Increased structural covariance in brain regions for number processing and memory in children with developmental dyscalculia. *Journal of Neuroscience Research, 100*(2), 522–536.

Mills, C., Tracey, D., Nash, S., & Gorkin III, R. (2023). Perceptions of a virtual reality sensory room for adults with neurodevelopmental disabilities. *Disability and Rehabilitation*, 1–10.

Morgan, H. (2019). Connections between sensory sensitivities in autism; the importance of sensory friendly environments for accessibility and increased quality of life for the neurodivergent autistic minority. *PSU McNair Scholars Online Journal, 13*(1), 11.

Morvan, P. (2008). Sensory experience and intentionalism. *Philosophy Compass, 3*, 685–702. https://doi.org/10.1111/J.1747-9991.2008.00148.X

Muraki, E. J., Speed, L. J., & Pexman, P. M. (2023). Insights into embodied cognition and mental imagery from aphantasia. *Nature Reviews Psychology, 2*(10), 591–605.

Mylonas, D., Caparos, S., & Davidoff, J. (2022). Augmenting a colour lexicon. *Humanities and Social Sciences Communications, 9*(1), 1–12.

Neufeld, J., Hederos Eriksson, L., Hammarsten, R., Lundin Remnelius, K., Tillmann, J., Isaksson, J., & Bölte, S. (2021). The impact of atypical sensory processing on adaptive functioning within and beyond autism: The role of familial factors. *Autism, 25*(8), 2341–2355.

O'Dowd, A., Hirst, R. J., Setti, A., Kenny, R. A., & Newell, F. N. (2022). Self-reported sensory decline in older adults is longitudinally associated with both modality-general and modality-specific factors. *Innovation in Aging, 6*(7), igac069.

Pacolet, J. (2022). Creativity in cyborgs: Mind, body and technology. *Psychoanalysis and the Mind-Body Problem*, 158.

Padfield, N., Zabalza, J., Zhao, H., Masero, V., & Ren, J. (2019). EEG-based brain-computer interfaces using motor-imagery: Techniques and challenges. *Sensors, 19*(6), 1423.

Pampaliari, S., Panagiotidis, P., & Sapounidis, T. (2023). Robot assisted language learning: Applications of educational activities with duckycode robot in the English classroom. In *ICERI2023 Proceedings* (pp. 2413–2421). IATED.

Peel, H. J., & Chouinard, P. A. (2023). A review of the impairments, preserved visual functions, and neuropathology in 21 patients with visual form agnosia—A unique defect with line drawings. *Neuropsychologia*, 108666.

Proulx, M., Brown, D., Pasqualotto, A., & Meijer, P. (2014). Multisensory perceptual learning and sensory substitution. *Neuroscience & Biobehavioral Reviews, 41*, 16–25. https://doi.org/10.1016/j.neubiorev.2012.11.017

Price, E., Tang, R., Kadri, H., & Gkatzionis, K. (2019). Sensory analysis of honey using Flash profile: A cross-cultural comparison of Greek and Chinese panels. *Journal of Sensory Studies*. https://doi.org/10.1111/JOSS.12494

Ptak, R., & Lazeyras, F. (2019). Functional connectivity and the failure to retrieve meaning from shape in visual object agnosia. *Brain and Cognition, 131*, 94–101.

Race, L., El-Amin, K., Anoke, S., Hayward, A., James, A., Hurst, A., ... & Mershon, T. (2022, October). Understanding design preferences for sensory-sensitive earcons with neurodivergent individuals. In *Proceedings of the 24th international ACM SIGACCESS conference on computers and accessibility* (pp. 1–4).

Rathee, D., & Mann, S. (2022, December). Daltonizer: A CNN-based framework for monochromatic and dichromatic color-blindness. In *2022 4th international conference on artificial intelligence and speech technology* (AIST) (pp. 1–5). IEEE.

Reitz, S. M., Scaffa, M. E., & Dorsey, J. (2020). Occupational therapy in the promotion of health and well-being. *American Journal of Occupational Therapy, 74*(3).

Rhodus, E. K., Hunter, E. G., Rowles, G. D., Bardach, S. H., Parsons, K., Barber, J., ... & Jicha, G. A. (2022). Sensory processing abnormalities in community-dwelling older adults with cognitive impairment: A mixed methods study. *Gerontology and Geriatric Medicine, 8*, 23337214211068290.

Riedel, A., Maier, S., Wenzler, K., Feige, B., Tebartz van Elst, L., Bölte, S., & Neufeld, J. (2020). A case of co-occurring synesthesia, autism, prodigious talent and strong structural brain connectivity. *BMC Psychiatry, 20*(1), 1–11.

Robotham, R. J., & Starrfelt, R. (2018). Tests of whole upright face processing in prosopagnosia: A literature review. *Neuropsychologia, 121*, 106–121.

Rosler, N., & Press-Barnathan, G. (2023). Cultural sanctions and ontological (in) security: Operationalisation in the context of mega-events. *Cambridge Review of International Affairs, 36*(5), 720–744.

Rossion, B. (2022). Twenty years of investigation with the case of prosopagnosia PS to understand human face identity recognition. Part II: Neural Basis. *Neuropsychologia, 173*, 108279.

Rudenko, S., & Haahr, M. (2023, October). Synaesthesia Gallery AR: Journey through the senses—Using augmented reality for education. In *Joint International conference on serious games* (pp. 211–217). Springer Nature Switzerland.

Rynkiewicz, A., & Łucka, I. (2018). Autism spectrum disorder (ASD) in girls. Co-occurring psychopathology. Sex differences in clinical manifestation. *Psychiatria Polska, 52*(4), 629–639.

Sabot, P. (2023). Michel Foucault in the 1950s: Beyond psychology towards radical ontology. *Theory, Culture & Society, 40*(1–2), 57–70.

Salehpour, F., Mahmoudi, J., Kamari, F., Sadigh-Eteghad, S., Rasta, S. H., & Hamblin, M. R. (2018). Brain photobiomodulation therapy: A narrative review. *Molecular Neurobiology, 55*, 6601–6636.

Saltagi, A. K., Saltagi, M. Z., Nag, A. K., Wu, A. W., Higgins, T. S., Knisely, A., ... & Illing, E. A. (2021). Diagnosis of anosmia and hyposmia: A systematic review. *Allergy & Rhinology, 12*, 21526567211026568.

Saniasiaya, J., & Prepageran, N. (2021). Impact of olfactory dysfunction on quality of life in coronavirus disease 2019 patients: A systematic review. *The Journal of Laryngology & Otology, 135*(11), 947–952.

Santos, F. H., Ribeiro, F. S., Dias-Piovezana, A. L., Primi, C., Dowker, A., & von Aster, M. (2022). Discerning developmental dyscalculia and neurodevelopmental models of numerical cognition in a disadvantaged educational context. *Brain Sciences, 12*(5), 653.

Schablitzky, S., & Pause, B. M. (2014). Sadness might isolate you in a non-smelling world: Olfactory perception and depression. *Frontiers in Psychology, 5*, 45.

Schoen, S. A., Lane, S. J., Mailloux, Z., May-Benson, T., Parham, L. D., Smith Roley, S., & Schaaf, R. C. (2019). A systematic review of Ayres sensory integration intervention for children with autism. *Autism Research, 12*(1), 6–19.

Schneider, M. L., Moore, C. F., Gajewski, L. L., Laughlin, N. K., Larson, J. A., Gay, C. L., ... & DeJesus, O. T. (2007). Sensory processing disorders in a nonhuman primate model: evidence for occupational therapy practice. *The American Journal of Occupational Therapy, 61*(2), 247–253.

Schwartzman, R., & Knowles, C. (2022). Expanding accessibility: Sensory sensitive programming for museums. *Curator: The Museum Journal, 65*(1), 95–116.

Sechrest, E. R., Chmelik, K., Tan, W. D., & Deng, W. T. (2023). Blue cone monochromacy and gene therapy. *Vision Research, 208*, 108221.

Semin, G. R., DePhillips, M., & Gomes, N. (2023). Investigating inattentional blindness through the lens of fear chemosignals. *Psychological Science*, 09567976231213572.

Simon, G. B., Abraham, F. A., & Melamed, S. (2004). Pingelapese achromatopsia: Correlation between paradoxical pupillary response and clinical features. *British Journal of Ophthalmology, 88*(2), 223–225.

Shaw, C., Churchill, V., Curtain, S., Davies, A., Davis, B., Kalenderidis, Z., ... & Thompson, G. A. (2022). Lived experience perspectives on ableism within and beyond music therapists' professional identities. *Music Therapy Perspectives, 40*(2), 143–151.

Sierzputowski, K. (2017, October 9). Collaborative acrylic paintings aim to visually map the perceptual experiences of synaesthesia. *Colossal.* https://www.thiscolossal.com/2017/10/collaborative-synesthesia-paintings/

Silva-Robles, F., Lopez, F., Duff, J., & Anderson, C. (2022). Eliciting associated motion constructions in two Zapotec languages. *Semantic Fieldwork Methods, 4*(3), 1–51.

Sindhuja, D. S. V., Bhateja, S., Sharma, M., & Arora, G. S. (2022). The untold saga of chromotherapy in dentistry. *Journal of Family Medicine and Primary Care, 11*(2), 453.

Sohrabi, Y., Rahimian, F., Soleimani, E., & Hassanipour, S. (2023). Low-level occupational exposure to BTEX and dyschromatopsia: A systematic review and meta-analysis. *International Journal of Occupational Safety and Ergonomics*, 1–11.

Song, Y. (2010). Art in nature and schools: Nils-Udo. *The Journal of Aesthetic Education, 44*, 108–196. https://doi.org/10.1353/JAE.2010.0003

Song, S. Y., Kim, Y. J., Lee, E. J., Yeh, S. W., Park, J. H., & Park, Y. G. (2023). Wintertime sea surface temperature variability modulated by Arctic Oscillation in the northwestern part of the East/Japan Sea and its relationship with marine heatwaves. *Frontiers in Marine Science.*

Spence, C. (2020). Scenting the anosmic cube: On the use of ambient scent in the context of the art gallery or museum. *i-Perception, 11*(6), 2041669520966628.

Stallings, J. W. (2022). *Special interests in art therapy with autistic people: A neurodiversity-positive approach to empower and engage participants*. Jessica Kingsley Publishers.

Stevenson, L. J. (2019). *Neither contradictory nor coincidental* (Doctoral dissertation, UC Santa Cruz).

Stokes, D. R. (2023). How radical is perceptual malleability? A reply to commentators. *Philosophy and the Mind Sciences*, 4.

Stumps, A., Saad, E., Rothlein, D., Verfaellie, M., & DeGutis, J. (2020). Characterizing developmental prosopagnosia beyond face perception: Impaired recollection but intact familiarity recognition. *Cortex, 130*, 64–77.

Suhaimi, N. S., Mountstephens, J., & Teo, J. (2020). EEG-based emotion recognition: A state-of-the-art review of current trends and opportunities. *Computational Intelligence and Neuroscience.*

Sultan, U., Axell, C., & Hallström, J. (2023). Bringing girls and women into STEM?: Girls' technological activities and conceptions when participating in an all-girl technology camp. *International Journal of Technology and Design Education*, 1–25.

Swathi, B., Koushalya, R., Roshan, J. V., & Gowtham, M. N. (2020). Color blindness algorithm comparison for developing an android application. *International Research Journal of Engineering and Technology (IRJET).*

Syma, C. (2019). Invisible disabilities: Perceptions and barriers to reasonable accommodations in the workplace. *Library Management, 40*(1/2), 113–120.

Tali, K. (2022). Contrastive analysis of English, Russian, and Hebrew color idioms. *European Journal of Literature, Language and Linguistics Studies, 6*(2).

Tamura, K., & Okamoto, T. (2023). Odor descriptive ratings can predict some odor-color associations in different color features of hue or lightness. *PeerJ, 11*, e15251.

Tosserams, A., Weerdesteyn, V., Bal, T., Bloem, B. R., Solis-Escalante, T., & Nonnekes, J. (2022). Cortical correlates of gait compensation strategies in Parkinson disease. *Annals of Neurology, 91*(3), 329–341.

Trauth, E. M. (2023). An overview of the individual differences theory of gender and IT. *Handbook of Gender and Technology: Environment, Identity, Individual*, 22.

Tromans, S., Chester, V., Gemegah, E., Roberts, K., Morgan, Z., Yao, G. L., & Brugha, T. (2021). Autism identification across ethnic groups: A narrative review. *Advances in Autism, 7*(3), 241–255.

Vázquez-Guardado, A., Yang, Y., Bandodkar, A. J., & Rogers, J. A. (2020). Recent advances in neurotechnologies with broad potential for neuroscience research. *Nature Neuroscience, 23*(12), 1522–1536.

Vinogradov, S. (2023). Harnessing neuroplasticity. *Psychiatry Research, 330*, 115607.

Wallace, R. (2007). Culture and inattentional blindness: A global workspace perspective. *Journal of Theoretical Biology, 245*(2), 378–390. https://doi.org/10.1016/J.JTBI.2006.10.006

Ward, R. J., Jjunju, F. P. M., Griffith, E. J., Wuerger, S. M., & Marshall, A. (2020). Artificial odour-vision syneasthesia via olfactory sensory argumentation. *IEEE Sensors Journal, 21*(5), 6784–6792.

Wapner, W., Judd, T., & Gardner, H. (1978). Visual agnosia in an artist. *Cortex, 14*(3), 343–364.

Watanabe, H., Ikeda, M., & Mori, E. (2020). Non-fluent/agrammatic variant of primary progressive aphasia with generalized auditory agnosia. *Frontiers in Neurology, 11*, 519.

Wei, Y., Zhang, Y., Wang, Y., & Liu, C. (2023). A study of the emotional impact of interior lighting color in rural bed and breakfast space design. *Buildings, 13*(10), 2537.

Widodo, S. A., Prihatiningsih, A., & Taufiq, I. (2021). Single subject research: Use of interactive video in children with developmental disabilities with dyscalculia to introduce natural numbers. *Participatory Educational Research, 8*(2), 94–108.

Wilcockson, T. D., Burns, E. J., Xia, B., Tree, J., & Crawford, T. J. (2020). Atypically heterogeneous vertical first fixations to faces in a case series of people with developmental prosopagnosia. *Visual Cognition, 28*(4), 311–323.

Wu, M., & Gingrich, O. (2020). *A feast for the eyes: Visualising flavour to vision synesthesia.*

Yan, Y., Yang, Y., Ando, M., Liu, X., & Kambara, T. (2021). Multisensory connections of novel linguistic stimuli in japanese as a native language and referential tastes. *European Journal of Investigation in Health, Psychology and Education, 11*(3), 999–1010.

Zhang, R., Zhang, X., Chen, Y., & Song, W. (2021). Current perception threshold testing in chronic ankle instability. *BMC Musculoskeletal Disorders, 22*(1), 453.

Zhou, Y., Meng, X., Wang, J., Mo, X., Jiang, S., Dai, C., & Liu, M. (2023). Daily peer relationships and academic achievement among college students: A social network analysis based on behavioral big data. *Sustainability, 15*(22), 15762.

CHAPTER 4

Diagnostics in Neurotribe Identification: Case Studies

This chapter navigates the complex world of neurodiversity in relation to artistic expression and perception. It begins by establishing a theoretical foundation, exploring enactivism and phenomenology to understand how sensory experiences are shaped by cognitive processes and personal realities. This foundation is brought to life through case studies and interviews demonstrating these concepts in practical scenarios. The chapter then delves into Sensory Processing Sensitivity (SPS), exploring how it affects aesthetic responses and is expressed in artistic interactions. Examining the intricate link between aesthetic preferences, emotional responses, and their underlying neurological mechanisms, offers insights gained from interviews around emotional reactions to art. Diagnostic approaches, such as the use of art diaries and pattern identification in artistic preferences, are discussed to understand sensory profiles better. The chapter also addresses the origins and impacts of sensory biases from infancy to adulthood, highlighting the diversity in artistic preferences among neurodivergent individuals, especially focusing on autism and sensory processing disorders. It explores the role of technology in aiding diagnosis and profiling, including AI-driven approaches and app-based neurotype tracking. The complex interplay of sensory experiences, such as synesthesia, and their influence on art perception, are also covered. Concluding with future directions, the chapter points to

J. Hutson et al., *Sensemaking and Neuroaesthetics*,
https://doi.org/10.1007/978-3-031-58045-1_4

emerging research areas like personalized quizzes and reduced chromoscapes, signaling new opportunities in the study of neurodiversity and art.

4.1 Foundations of Neurotribe Identification: Enactivism and Phenomenology

Millennia before the advent of psychoanalysis or psychotherapy, a fundamental inquiry into self-identity and awareness was espoused by the insightful humanist dictum of ancient Greek philosopher Socrates, γνῶθι σεαυτόν (*gnōthi seauton*) or "Know thyself" (Bowery, 2007). This core theme of the chapter urges individuals to understand and identify with their unique neurological makeup or neurotribe and their particular sensory profiles. A neurotribe, or neurological community, refers to a group of individuals who share similar neurological or cognitive patterns, often used in the context of neurodiversity to describe groups like the autistic, dyslexic, ADHD, dyspraxic, and OCD communities (Silberman, 2017). These societies are not exclusive and sometimes overlap; for instance, it is common for someone on the autism spectrum to also have ADHD or dyslexia as a co-occurrence (Lin et al., 2023). On the other hand, sensory profiles encompass the specific ways individuals perceive and process sensory information, including preferences and sensitivities across the senses such as visual, auditory, tactile, and olfactory (Licciardi & Brown, 2023). Taken together, knowing one's neurological diagnoses and understanding how sensory profiles impact environmental experiences, empower individuals to more productively engage in the world and promote better health and well-being (Hitch et al., 2023; Merten et al., 2022; Wendin et al., 2023).

While the benefits of understanding sensory profiles in promoting better quality of life have been well-documented, the "hidden curriculum" of society, or the unofficial, unintended lessons, values, and perspectives that students learn in school and other institutions, typically promotes homogeneity and adherence to a one-size-fits-all approach (Ahvan et al., 2022; Cid Rodriguez et al., 2022; Cunliffe et al., 2022). Such hegemonic systems often neglect the diversity of neurotypes, defined as variations in the neurocognitive functioning of individuals, and fails to acknowledge that not all environments are compatible with all neurological makeups (Evans et al., 2023; Stanek & Mattson, 2023). As a result, individuals may not be taught to recognize stimuli that are anxiety-inducing or uncomfortable for them, nor are they typically instructed in self-regulation methods or how to seek out environments that accommodate their neurotype (Bernett, 2022; Hutson & Hutson, 2023a).

This chapter, therefore, aims to address this gap by first considering the interaction between our environment and our brain in creating our experience of reality. It then explores different sensory profiles and diagnostic criteria to help individuals identify their neurotribe and understand their sensory processing. Through interviews and case studies, this chapter will offer insights into various questionnaires and diagnostic strategies that can help determine one's neurotype. The ultimate goal is to provide strategies for anyone to reflect on their sensory processing of the world, leading to self-awareness and identification of experiences that result in aesthetically pleasing or displeasing outcomes, regardless of being neurotypical or neurodivergent. By understanding the foundations of neurological group identification and how individuals process their experiences of the world, one can better navigate and appreciate the richness of the sensory world and the diversity of experiences it offers.

In order to assist in identifying diverse neurotribes, understanding how humans interpret knowledge about their worlds, or knowing *how* we know what we know is critical. As such, the epistemological frameworks of *enactivism* and *phenomenology* offer essential frameworks for understanding sense experience and interpreting external stimuli. These two frameworks should be considered because they approach cognition from a holistic and embodied perspective, which contribute significantly to understanding neurotribes, particularly in the context of experiencing and creating art. These frameworks help in interpreting how people with different neurological makeups perceive and interact with their environments and artworks. Enactivism, for instance, emphasizes the dynamic interaction between an organism and its environment, suggesting cognition is an active process of engagement and sensemaking (Meyer & Brancazio, 2022). Phenomenology, on the other hand, focuses on the first-person experience and the essential structures of consciousness, providing a deep understanding of how things are meaningfully experienced (Gallagher, 2022). Together, they offer a more comprehensive view of human experience, highlighting the importance of subjective perception, embodiment, and the active role of individuals in shaping their understanding of the world. Considering both frameworks allows for a richer and more nuanced appreciation of the complexity of sensory experience and cognitive processes.

Enactivism, particularly, posits that *sensemaking* emerges from relational processes distributed across the brain-body-environment system, suggesting that our understanding of reality is deeply rooted in our active engagement with the world around us (Sepúlveda-Pedro, 2023). Autopoietic enactivists and phenomenologists like Merleau-Ponty (1908–1961) explore sensemaking with a focus on sentient beings, emphasizing the relational and embodied aspects of cognition (Christias, 2023). The predictive capabilities of the brain also play a crucial role in interpreting sensory information, employing memory of past experiences to anticipate and make sense of incoming stimuli (Pezzulo et al., 2022). This continuous interpretation and prediction mechanism are fundamental to our perception and understanding of reality. The approach to understanding sensemaking with enactivism can serve as a flexible resource for researchers exploring the unfolding of first-person experiences and its meanings, focusing on the embodied and dynamic interaction between the organism and the environment (Di Paolo, 2023). Furthermore, this perspective suggests that art perception and creation are active processes of sensemaking, involving the whole of engagement with art and its context (Jelić et al., 2016; Bar, 2020).

Along with the first-person approach of enactivism to understanding sensemaking, phenomenology, as the interpretive study of human experience, aims to expand and examine human situations, events, meanings, and experiences in daily life (Ayton, 2023). It provides a foundation for understanding how individuals experience and interpret the world outside of themselves through interactions with it, focusing on the subjective and interpretative nature of cognition (Kyzar & Denfield, 2023). Enactivism and Sartrean phenomenology together offer a broader analysis of consciousness and cognition, highlighting the interconnectedness of body, mind, and world (Pawlett-Jackson, 2022). It places a strong emphasis on intentionality, the lived body, and the pre-reflective level of experience. For neurotribes, this means that their unique sensory and cognitive processing styles are recognized and valued as legitimate ways of experiencing art and the environment. Phenomenology helps understand how individuals with different neurotypes might perceive and interpret art in unique ways that are deeply subjective and tied to their particular embodied experience (Banks, 2022; Pantazakos & Vanaken, 2023).

Together, enactivism and phenomenology offer a comprehensive understanding of how neurotribes experience and interact with art and their environments. They recognize the diversity of sensory and cognitive experiences, advocating for art and environmental designs that are inclusive and responsive to the needs of different neurotypes. These frameworks are crucial for developing therapeutic art programs, inclusive museum practices, and educational strategies that respect and accommodate the diversity of sensory experiences and cognitive styles within neurotribes. However, the approaches should also be considered for diagnostic purposes in discussing sensory profiles. The study conducted will be reviewed within these frameworks, focusing on the individuality of sense-making and the experiences of different neurological communities. The investigation aims to balance the recognition of the unique ways individuals experience the world with the identification of common patterns across neurotypes. This balance will provide a foundation for creating tailored experiences that cater to various sensory processing conditions. Essentially, while acknowledging that each person interacts with the world and aesthetic experiences uniquely, the study seeks to discern shared experiences and preferences that can inform the development of customized interventions and supports.

4.2 Sensory Processing and Aesthetic Responses Study

Sensory processing and aesthetic responses are intimately connected, with various case studies and research providing insights into the ways individuals interact with and respond to art. Aesthetic experience in viewing art depends on both sensory processing and conceptual processing, where specific brain regions correlate with the perception of motion and the pleasantness of the viewed art (Hagtvedt, 2022). This highlights the importance of understanding *sensory processing sensitivity* (SPS) and its influence on aesthetic experiences (Sadeghzadeh et al., 2023). People with heightened sensory processing sensitivity may experience enhanced emotional reactions to art, including physical sensations like goosebumps or a lump in their throat. Tihanyi et al. reported on the neuropsychophysiology of tingling in their 2018 study using electric stimulation. These reactions are influenced by both pleasurable and unpleasurable art, with studies showing similar brain activity patterns when experiencing positive emotions like love or awe during the viewing of aesthetically pleasing art (Bolwerk et al., 2014). Conversely, Vessel et al. (2012) reported that unpleasing art activates brain regions associated with negative emotions, such as disgust or fear.

Understanding how emotions and the psychological effects of stress on the autonomic nervous system play into aesthetic responses is crucial. The right anterior insula, for instance, is consistently activated across all sensory modalities during aesthetic appraisals, Brown et al. noted (2011) and Thielen et al. (2023) echoed, suggesting a primary role in assessing the valence of perceived objects, not just artworks. As Jacobi et al. (2022) relate in their four-country study, this indicates the integral role of emotions and the stress responses of the body in shaping our aesthetic experiences. In terms of post-traumatic stress disorder (PTSD) and its implications on Broca's area of the brain (Fig. 4.1), Sheehan-Klein (2024) report that art assists in self-regulation and treatment for traumatic brain injury and PTSD. The therapeutic use of art activates various brain areas and supports emotional healing, providing an outlet for expression when verbal communication might be impaired (Malik, 2022). Moreover, research into the developmental aspects of aesthetic preferences, such as the preference for landscape paintings rooted in sensory biases from infancy, shows that our sensory experiences and biases shape our aesthetic preferences throughout our lives (Kliem et al., 2022). This extends to neurodivergent populations, such as autistic individuals who may have distinct preferences for more realistic and representational art, reflecting their unique ways of processing and perceiving the world (Miller & Hübner, 2023).

Research into synesthesia proper has also provided a wealth of information on the neural and cognitive basis of this unique sensory phenomenon, informing hypotheses and methodologies for studying sensory processing and aesthetic responses (Cytowic, 2002; Hubbard, 2007; Ward, 2013). These studies have employed various research questions, hypotheses, and methods to gather data on synesthetic experiences and their implications for aesthetic responses. Synesthesia research often focused on the neural basis of subjective experiences, aiming to understand how cross-modal associations impact perception and cognition in both healthy and pathological brains (Neckář & Bob, 2014; Simner et al., 2006; Weiss et al., 2005). These past studies revealed synesthesia involves anomalous activation of color-selective areas and hyper-binding mediated by the parietal cortex, suggesting unique mechanisms of sensory integration (Hubbard, 2007). One critical methodology in synesthesia research is

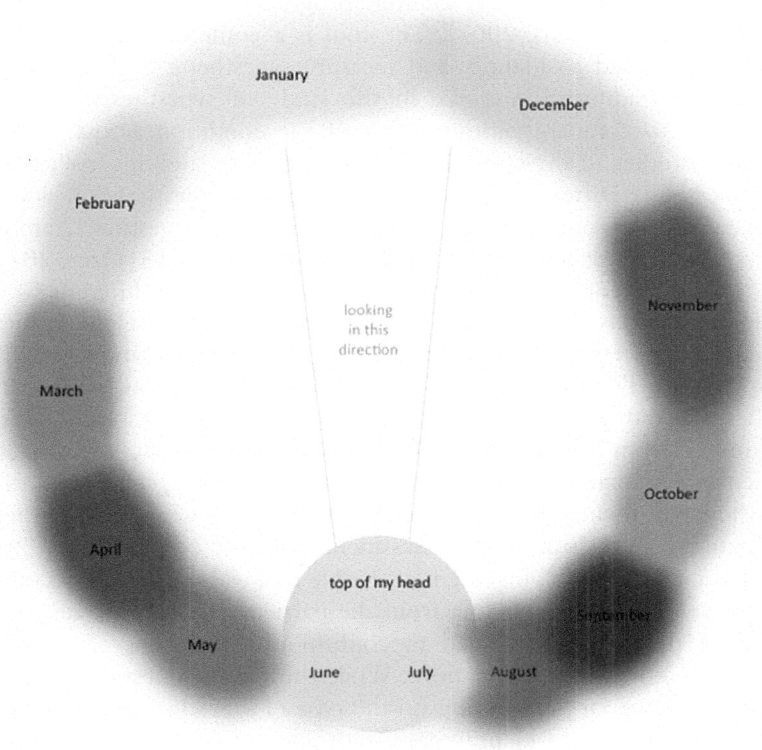

Fig. 4.1 Motor and sensory regions of the cerebral cortex. Blausen.com staff. (2014). Medical gallery of Blausen Medical 2014. *WikiJournal of Medicine*, *1*(2). https://doi.org/10.15347/wjm/2014.010. ISSN 2002-4436. (*Source* From Wikimedia Commons, licensed under CC-BY 3)

the use of double-blind studies to limit biases and promote replicability. For instance, a study on grapheme-color synesthesia utilized a double-blind design to assess visual imagery usage in synesthetes, providing insights into the cognitive and neural underpinnings of the condition (Brang & Ahn, 2019).

The Synesthesia Battery is another tool used in research to provide a standardized test battery for comparing and contrasting synesthetic subjects, thereby accelerating scientific progress in understanding synesthesia (Eagleman et al., 2007). The tool is a comprehensive standardized test designed to identify and measure synesthetic experiences and has been a traditional resource in the study of synesthesia, allowing researchers to compare various forms of the condition across individuals and studies. Developed by David Eagleman and his colleagues, the Synesthesia Battery assesses different types of synesthesia, including grapheme-color, sound-color, and lexical-gustatory synesthesia, among others (Bankieris & Simner, 2014; Meier & Rothen, 2013; Özer, 2019). The tool is structured into several sections, each aimed at different synesthetic associations. Participants are typically asked a series of questions designed to ascertain whether they experience synesthesia, and if so, what form it takes. For example, in the grapheme-color section, participants might be asked to choose the color they associate with specific letters or numbers from a color palette. The tool attempts consistency by asking participants to match colors to graphemes multiple times, checking for the consistency of their responses, which is a hallmark of genuine synesthetic experiences. The design of the assessment is mostly online (https://synesthete.ircn.jp/pretest), making it accessible to a wide audience. Its effectiveness relies on self-reporting from the participants, and the repetitive nature of the tests is intended to confirm the consistency and specificity of the synesthetic associations. While the specific questions asked in each section can vary depending on the type of synesthesia being tested, the overall intent is to identify consistent and involuntary cross-sensory experiences characteristic of synesthetic individuals.

While the Synesthesia Battery has been instrumental in advancing research into synesthesia, several limitations and criticisms have been noted regarding its reliability and methodology. One of the main criticisms is that most differences claimed for synesthetes are unsupported due to low statistical power, statistical errors, and methodological limitations in the research (Hupé & Dojat, 2015). Additionally, while the Synesthesia Battery is a valid methodology for assessing synesthesia, with a prevalence of 1.2% in the general population, concerns have been raised about the test–retest procedure and the reliance on online battery assessments for understanding the nature of synesthesia (Carmichael et al., 2015). The "test–retest" method, where individuals are tested multiple times to confirm the consistency of their synesthetic experiences, may not capture the full complexity and variability of synesthetic perceptions. Furthermore, current neurocognitive models of synesthesia primarily focus on modality-specific brain regions for perception, often neglecting crucial psychophysical characteristics documented in behavioral research (Chiou & Rich, 2014). This could lead to an incomplete understanding of the condition and its manifestations across different individuals. Other limitations include the fact that alternate coupling within the same network of active regions leads to differences in subjective experience in grapheme-color synesthesia (van Leeuwen et al., 2011). This suggests individual variations in brain connectivity and structure might significantly influence synesthetic experiences, making it challenging to generalize findings across all synesthetes.

Evolving research has indicated that different types of synesthesia involve distinct neural mechanisms, with each synesthete having unique experiences tied to their specific inducer and concurrent associations (Hauw et al., 2023). Understanding these individual differences is vital for appreciating the diversity of synesthetic experiences and their impact on aesthetic responses. For example, grapheme-color synesthesia is characterized by enhanced cortical excitability in the primary visual cortex, influencing conscious awareness of color photisms (Andreev et al., 2023). In understanding the broader implications of synesthesia for aesthetic experiences, these studies illustrate the synesthetic experiences can alter brain networks and connectivity, revealing the interplay between semantics and sensory input (Aoki et al., 2023; Churchill & Wilde, 2022; Lebeau & Richer, 2022). This has significant ramifications for the way synesthetes encounter art, as they might possess intensified emotional and sensory reactions to specific stimuli, shaping their overall aesthetic encounter.

Previous methodologies and research have significantly informed the current study on neuroarts and neuroaesthetics, particularly in understanding the sensory processing conditions and aesthetic responses of individuals with various neurotribes. The research highlights the importance of understanding commonalities between similar neural profiles, such as sensory processing disorders, and examining how individuals perceive and interact with their environment through various senses. Specific stimuli like touch, smell, sound, or images can have a unique impact on an individual, and there may be a connection between sensory preferences and visual stimuli. These findings play a pivotal role in aiding individuals in recognizing their triggers and acquiring strategies to regulate them, ultimately fostering enhanced self-awareness and self-actualization. Moreover, the digital realm, with its potential for customization and regulation, presents distinctive prospects for tailoring experiences to meet the diverse requirements of neurodivergent individuals. This encompasses the development of tools and applications that enable users to encounter the world through different sensory conditions, including synesthesia, variations in sensory processing times, or color blindness. Profound comprehension of these conditions and preferences is imperative for crafting more accommodating and effective interventions.

The present investigation is dedicated to enhancing our comprehension of neurodiversity, with the objective of aiding individuals in recognizing their specific neurological profiles and gaining insight into how they process sensory data and cognition. This research endeavors to transcend standardized, uniform solutions and instead, acknowledges and appreciates the intricate diversity inherent in human neurodiversity. Using a comprehensive approach that includes various neurodiverse groups, such as people with synesthesia, the study employs both surveys and one-on-one interviews to delve into the diverse experiences of these individuals (Hutson & Hutson, 2023b). The research questions for the study included:

1. *How do individuals with different neurotypes perceive and respond to art and aesthetic experiences?*
2. *What are the specific sensory preferences and aversions associated with different neurotypes?*
3. *How do neurodivergent individuals' art preferences compare to those of neurotypical individuals?*
4. *What are the commonalities and differences in aesthetic experiences across various sensory processing conditions, such as synesthesia or sensory processing disorders?*

These research questions aim to deepen the understanding of the complex interplay between neurodiversity, sensory processing, and aesthetic experiences, and guide the development of more empathetic and effective approaches in art therapy, education, and exhibition design. The data gathered from the study is further analyzed to develop new ways of determining one's neurotribe. Once individuals understand their neurotribe and sensory profile, along with the diagnostic criteria relevant to them, they can better manage their reactions to stimuli and create tailored experiences that suit their specific needs. This understanding allows for greater control over one's environment and interactions, leading to more personalized and comfortable experiences. The results from this study are essential, as they guide the discussions in the book and provide a base for examining specialized interventions and strategies that align with the varied sensory needs and preferences of the participants of the study.

In the study examining neurodiversity, a diverse demographic of 94 participants provided insights into a spectrum of neurotypes, underscoring the prevalence and variety within the population. The participants reported conditions including dyslexia (6.38%), dysgraphia (1.06%), ADHD (23.40%), PTSD (11.70%), Autism Spectrum Condition (ASC) (9.57%), and other unspecified neurotypes (12.77%). Notably, 19.15% of respondents identified as neurotypical (Table 1.1). The age distribution ranged from 18 to over 65, with the largest group being those aged 35–44 (26.56%). Gender identity was predominantly female (75%), with males (23.44%) and a smaller fraction identifying as genderqueer/genderfluid (1.56%).

Cultural identity was primarily non-Hispanic/Latinx (93.44%), while a variety of races and ethnic heritages were represented, with the majority being White/Caucasian (85.07%). Sexual orientation varied with the majority identifying as heterosexual/straight (72.31%), followed by bisexual (10.77%), and smaller percentages for other orientations. Educationally, participants were largely well-educated with degrees ranging from bachelor's (32.81%) to doctoral (10.94%), and the majority spoke English at home (87.50%). Physical disabilities were reported by a minority of participants, with 25% having various impairments, including visual and mobility disabilities.

The diversity in age, gender, cultural background, and education level reflects a broad cross-section of society, providing a comprehensive view of how various neurotypes are distributed across different demographics. Additionally, the presence of various physical disabilities among participants indicates the multifaceted nature of individual challenges and experiences (Lebrasseur et al., 2021). This rich demographic landscape sets the stage for a nuanced exploration of how different people perceive and interact with their environments, particularly in the context of art and aesthetics, emphasizing the need for personalized approaches in neuroarts and neuroaesthetics research and practice.

The subsequent sections of the survey focused on general experiences with art, as well as emotional, sensory, and crossover sensations of participants. To gain deeper insights into these experiences, participants were asked the following questions:

1. *How often do you visit museums or art galleries?*
2. *How often do you view art on desktops, tablets, or mobile devices for enjoyment?*
3. *How often do you experience strong sensory reactions (e.g., goosebumps, tingling, emotional overwhelm) when engaging with art?*
4. *If applicable, do you believe your neurodivergence influences the way you perceive or engage with art? (This question is voluntary and for those who identify as neurodivergent.)*
5. *Do you ever experience sensory "crossovers" or unusual associations between different senses, known as synesthesia? (e.g., hearing/seeing color or seeing sound/smell)*
6. *What sensations or perceptions do you experience with your synesthesia?*
7. *Do you ever associate numbers or letters with specific colors, tastes, or smells?*
8. *Do certain words or phrases evoke particular textures or temperatures for you?*
9. *Do you experience time as a spatial dimension, like a mental calendar or map?*
10. *How does synesthesia affect your daily life or your perception of the world?*
11. *Do you find yourself drawn to or distracted by certain types of art?*
12. *Do you find that your emotional reactions to art are more intense or vivid than other people's?*
13. *Is there anything else about your exper*ience with art that you would like to share?

In the study, participants reported varied levels of engagement with art both in physical settings like museums and galleries and through digital platforms. A significant number of respondents frequented museums and galleries, with the majority visiting at least occasionally and a considerable 37.50% frequently engaging five or more times a year. Digital art consumption was even more prevalent, with the majority (56.25%)

viewing art on devices daily, indicating a strong integration of art into daily life and a preference for digital mediums.

Regarding emotional and sensory reactions to art, most participants (64.06%) sometimes experienced strong reactions like goosebumps or emotional overwhelm, while a significant portion felt these reactions most of the time or always when engaging with art. This suggests a deep sensory and emotional engagement with art among many respondents. Furthermore, half of the neurodivergent participants acknowledged that their neurodivergence positively influenced their art experience, indicating a significant relationship between neurodiversity and art perception (Moraru, 2023).

The survey section on sensory "crossovers" or synesthesia revealed that a substantial number of participants experienced various forms of synesthetic sensations (Table 1.2). About 9.38% always experience synesthesia, and 15.63% often have such experiences, indicating a notable presence of synesthetic perception in the study's demographic. The reported synesthetic experiences varied widely, with 12.86% seeing colors when hearing sounds, and others tasting flavors in response to words or numbers, feeling tactile sensations from visual patterns, and associating days or letters with colors and emotions. Unique synesthetic experiences were also documented, including mirror-touch synesthesia and hearing pain, showcasing the broad spectrum and individuality of synesthetic perceptions. These experiences sometimes extend beyond sensory perception to influence emotions, creativity, and physical sensations, illustrating synesthesia's profound impact on overall experience. Some respondents noted the creative influence of synesthesia on their artistic expression, curating soundscapes, and visuals. Furthermore, a significant portion of participants associated numbers or letters with specific colors, tastes, or smells, with varying frequencies and intensities. The survey also revealed diverse perceptions of time as a spatial dimension, indicating a complex interaction between synesthesia and temporal cognition.

In terms of the impact of synesthesia on daily life, responses indicated both positive and negative effects. Some participants reported enhanced memory, vivid experiences, and creative thinking, while others experienced confusion, overstimulation, or avoidance of certain stimuli. Despite the challenges, a notable percentage reported an enhanced appreciation of music, art, and other sensory experiences, highlighting the dual nature of synesthesia as both a unique sensory phenomenon and a significant factor in personal aesthetic and sensory experience. These findings underscore the complexity of synesthetic experiences and the need for further exploration into how they shape individual perceptions and interactions with the world (Hoffman et al., 2019). The survey data collected from the study on neurodiversity and sensory processing also provides comprehensive insights into the aggregate experiences of individuals with various neurological profiles and sensory conditions. A significant portion of participants actively engage with art both in physical spaces and digitally, with many experiencing intense emotional and sensory reactions to art.

From the aggregate data, it is clear that participants exhibit a wide range of interactions with and responses to art, influenced by their unique sensory processing conditions and neurotypes. For many, art engagement is frequent and deeply integrated into their daily lives, often enhanced or affected by conditions like synesthesia. The variety of synesthetic experiences reported, from seeing sounds to tasting words, underscores the diversity of sensory experiences present within the global population. Additionally, the impact of synesthesia on aspects like memory, creativity, and daily life experiences highlights its significance in shaping personal worldviews and interactions. The data also reveals the importance of considering individual differences in sensory processing and neurotype when creating and presenting art. As participants exhibit a broad spectrum of reactions and preferences, the need for personalized and inclusive art experiences becomes evident. This calls for a deeper understanding of how different neurotypes and sensory conditions influence aesthetic perception and engagement.

With these aggregate insights in hand, the study transitions to a more nuanced review of the data based on specific reported diagnoses, conditions, and synesthetic experiences. This next phase will delve into the individual differences and patterns within subgroups of the participant population, exploring how distinct neurotypes, such as ADHD, autism, or dyslexia, and different forms of synesthesia, affect individual experiences with and responses to art. By dissecting the data in this way, the study aims to uncover more detailed and tailored insights that can inform targeted strategies and interventions in art therapy, museum design, and broader neuroaesthetic applications. The goal is to provide a more comprehensive and empathetic understanding of the diverse sensory worlds inhabited by individuals, enriching the collective appreciation of art and aesthetic experiences.

4.3 Sensory Profiles and Aesthetic Preferences Case Studies

4.3.1 Neurological Conditions

4.3.1.1 Neurotypical Population

The various case studies from the survey provide insight into the variety of crossover experiences of different neurotribes. The section opens with a baseline understanding of the sensory experiences of neurotypical respondents. This foundational understanding is crucial to amalgamating with the most to least common neurodivergent conditions reported. The goal of the study is to uncover connections between neurological conditions and various reported sensory experiences, with a particular focus on crossover experiences like synesthesia, which blend senses in unique ways. The initial focus on neurotypical sensory experiences provides a standard against which to measure and understand the variances found in neurodivergent populations. The survey data has illuminated a range of interactions with art, with some individuals deeply integrated into artistic engagement, while others have distinct, sometimes intense sensory reactions or unique synesthetic experiences. Understanding these baseline experiences is key to appreciating the full spectrum of sensory and aesthetic responses.

The neurotypical respondents present with a diverse demographic profile, contributing to the baseline understanding of sensory experiences and aesthetic preferences. In terms of age, the group spans from young adults to those over 65, with the most substantial representations in the 25–34 and 45–54 age brackets, each comprising over a fifth of the participants. This age diversity ensures a wide range of life experiences and perspectives on art and sensory experiences. Gender identity among the neurotypical respondents is predominantly female, making up two-thirds of the group, while males constitute one-third. This gender distribution provides insights into how different genders might engage with and respond to art and sensory stimuli. The majority of these respondents are not Hispanic/Latinx, with a small minority affirming this cultural identity, and the racial heritage is overwhelmingly White/Caucasian with a few individuals identifying as Black or African-American or other unspecified ethnicities. Such cultural backgrounds could influence their sensory processing and aesthetic experiences.

Sexual orientation is primarily heterosexual/straight, with a small representation of bisexual individuals and those preferring not to disclose their orientation. Educationally, the group is well-educated, with a significant number holding bachelor's or master's degrees, indicating a potentially higher level of engagement or understanding of the arts and culture. The primary language spoken at home is predominantly English, with some respondents speaking Portuguese, Arabic, or other languages, possibly affecting their cultural and sensory experiences. Physical disabilities are notably absent in the group, except for one individual who identifies being overweight as a disability. This lack of physical disabilities might influence the types of sensory experiences and preferences reported, as physical conditions can significantly affect how individuals interact with and perceive their environments.

The neurotypical population in the study reported varied levels of engagement with art, revealing insights into their frequency of visits to museums or art galleries, digital consumption of art, and sensory reactions to art. A significant 55.56% of the neurotypical respondents frequently visit museums or art galleries five or more times a year, indicating a robust engagement with physical art spaces. Another 27.78% visit occasionally, 2–4 times a year, while 11.11% visit rarely, once a year or less. Only a small minority, 5.56%, reported never visiting museums or art galleries. This pattern suggests that a majority of the neurotypical individuals are quite active in seeking out physical art experiences, showcasing a substantial interest and engagement in cultural and aesthetic environments (Adams et al., 2022). Moreover, when it comes to viewing art on desktops, tablets, or mobile devices, the majority (55.56%) of the neurotypical respondents engage with art daily, highlighting a significant preference for integrating art into their everyday digital experiences. Others view art 2–3 times a week (16.67%) or once a week (16.67%), with a smaller group engaging 4–6 times a week (11.11%). The high daily engagement indicates that digital platforms are a prominent medium for art consumption among the neurotypical population, possibly due to the convenience and accessibility of these platforms (Pesce et al., 2019).

In terms of sensory reactions to art, a majority of the neurotypical respondents, 72.22%, sometimes experience strong sensory reactions such as goosebumps, tingling, or emotional overwhelm when engaging with art. About 16.67% feel these reactions about half of the time, while a smaller portion experience them most of the time (5.56%) or always (5.56%). These findings indicate that while strong sensory reactions to art are common among neurotypical individuals, the frequency and intensity of these reactions vary, suggesting a range of emotional and sensory engagement with art (Winner, 2019).

In the neurotypical population, experiences with synesthesia varied, with 50% never or rarely experiencing these unusual associations between different senses. However, a notable minority reported occasional to frequent synesthetic experiences. The types of synesthetic sensations reported include seeing specific colors when hearing certain sounds (16.67%), feeling tactile sensations when observing visual patterns (16.67%), and occasionally associating letters or numbers with particular personalities or emotions (5.56%). Some participants also occasionally visualize time as a spatial dimension, like a mental map or calendar (49.99%). When asked about the influence of synesthesia on daily life, participants reported a variety of effects. About 36.36% stated that synesthesia enhances their appreciation or understanding of music, art, or other sensory experiences. In contrast, 27.27% noted it enhances memory and recall capabilities. Some find that it makes everyday experiences more vivid and immersive (9.09%) or facilitates creative thinking (9.09%). However, a small number also reported confusion or overstimulation in certain environments (9.09%).

In terms of art preferences, respondents showed a diverse range of tastes. Realistic (27.78%) and mixed media (27.78%) art forms were the most popular, followed by abstract (16.67%) and digital art (16.67%). This suggests that neurotypical individuals have varied aesthetic preferences, with no single style dominating. Emotional reactions to art also vary, with a significant 61.11% reporting their reactions as more intense than others, indicating a deep emotional engagement with art. However, 27.78% feel their reactions are comparable to others, highlighting the subjective nature of art perception and emotional response. The findings from the neurotypical population provide a baseline understanding of the frequency and nature of sensory crossovers like synesthesia, as well as general art preferences and emotional reactions to art. This baseline will be invaluable as we move on to compare these experiences with those reported by neurodivergent individuals, particularly those with autism, to uncover potential patterns or differences in sensory processing and aesthetic response. The goal is to understand the unique ways individuals experience and interact with art, informed by their neurological and sensory profiles, to enhance our approach to accommodating a diverse range of sensory needs and preferences in art and design.

4.3.1.2 Autistic Population

After the neurotypical population, those that reported to have a diagnosis of autism, autism spectrum condition (ASC), or Asperger's syndrome represented the next largest population. This community often reports co-occurrences of other conditions, offering a complex tapestry of experiences and challenges (Bouzy et al., 2023). In considering the neurotypical baseline for sensory experiences in art, the standard is against other neurotypes to measure the sensory and aesthetic experiences of individuals, starting with those with autism and associated co-occurrences. The comparison serves as a crucial step in comprehending how various neurotypes may engage with and interpret art in distinct ways. It seeks to unveil any unique trends or predilections which manifest among individuals, with a particular focus on how neurodivergent conditions might impact sensory encounters, aesthetic inclinations, and potentially, synesthesia-like cross-sensory occurrences.

Understanding autism is crucial for contextualizing the subsequent findings and discussions. First, autism spectrum encompasses a range of neurodevelopmental conditions, including what was formerly diagnosed as Asperger's Syndrome (Woodbury-Smith & Volkmar, 2009). While both ASC and Asperger's involve social communication difficulties and restricted, repetitive behaviors and interests, the latter is typically characterized by no significant delays in language or cognitive development, distinguishing it from other forms of autism which might present with a broader range of symptoms including intellectual disabilities and more pronounced sensory sensitivities (Borisova et al., 2023; de Giambattista et al., 2019; Smith & Jones, 2020). As such, the American Psychiatric Association (APA) subsumed Asperger's syndrome under the broader category of ASC in 2013, though the term continues to be used informally by many (Hase et al., 2023). The shift reflects a broader understanding that the spectrum of autism is wide and varied, with different individuals requiring different levels of support. Additionally, the lines between these diagnoses are often blurred, and the lived experiences of individuals may not fit neatly into these categories resulting in common co-occurrences of other diagnoses, such as ADHD, dysgraphia, dyslexia, PTSD, etc. (Micai et al., 2023).

In fact, the study revealed several co-occurring conditions with those reporting cases of autism, emphasizing the complexity and diversity of experiences within this neurodivergent group. A significant percentage of participants with autism also reported having ADHD, with 41.82%

of individuals indicating this co-occurrence. This finding underscores the common overlap between autism and attention-deficit/hyperactivity disorder, reflecting the shared characteristics and challenges that individuals might face in sensory processing, attention, and behavior (Mattheisen et al., 2022). Dyslexia was reported by 5.45% of the autistic population, highlighting that some individuals navigate the combined challenges of autism spectrum conditions and difficulties related to reading, language, and processing written words. Dysgraphia, a condition affecting writing abilities, was noted by 1.82% of participants, suggesting an impact on fine motor skills and expression for a minority within this group. Post-traumatic stress disorder (PTSD) was present in 9.09% of individuals with autism, indicating that some may be dealing with the additional complexity of trauma and its effects on emotional regulation and sensory sensitivity. The presence of PTSD can further complicate the sensory and emotional experiences of individuals with autism, requiring particular attention and care in understanding and support (Golan et al., 2022).

The study also identified that 14.55% of participants labeled their condition specifically as ASC, while a slightly lower percentage (12.73%) identified simply as having "autism." A subset of 5.45% specified Asperger's syndrome as their condition, reflecting the previous diagnostic nomenclature and varied ways individuals understand and describe their experiences within the autism spectrum. Additionally, 7.27% of participants with autism indicated other co-occurring conditions or neurotypes, suggesting a range of experiences and diagnoses that might not fit neatly into the more commonly recognized categories. This "other" category reflects the individual variability and the need for nuanced understanding and support tailored to each person's unique set of conditions and experiences.

The autistic population in the study presents a demographic profile that offers some contrasts and parallels to the neurotypical group. In terms of age, the autistic respondents are relatively younger, with a notable concentration in the 25–34 age bracket (26.09%) and significant representation across other age groups, except for those over 65. This younger skew might reflect the age demographics of individuals more likely to participate in studies or could suggest specific age groups more inclined to self-identify or receive a diagnosis (Adams & McGuire, 2022). Gender identity among the autistic respondents is predominantly female, accounting for 78.26% of the group, a higher percentage compared to the neurotypical respondents. This could reflect the growing recognition and diagnosis of autism in females or possibly the demographic of those more likely to engage in such a study (Hutson & Hutson, 2023c). A small minority identify as genderqueer/genderfluid, adding to the diversity of gender experiences within the autistic population.

Cultural and racial demographics show that 95.65% of autistic individuals do not identify as Hispanic/Latinx, and 86.96% identify as White/Caucasian, similar to the neurotypical group. This consistency suggests that the cultural background does not markedly differ between the two groups in the context of this study. Sexual orientation among autistic individuals shows a diverse spectrum, with a significant portion identifying as heterosexual/straight (58.33%) and notable representations of bisexuality, pansexuality, and queer identity. This variety indicates a range of personal identities and experiences within the autistic community, perhaps reflecting the broader diversity recognized within neurodivergent populations (Cohen et al., 2022).

The education level among autistic participants is varied, with a substantial number holding bachelor's or higher degrees, indicating a well-educated group. The presence of individuals across different educational achievements, including those with associate and doctoral degrees, reflects the diversity of educational backgrounds and intellectual pursuits within the autistic community (Steinbrenner et al., 2022). English predominates as the primary language spoken at home, mirroring the neurotypical group. A small percentage of the autistic group speaks multiple languages at home, indicating linguistic diversity within the community. Physical disabilities are more reported among autistic individuals compared to the neurotypical population, with a notable presence of mobility or physical disabilities, visual impairments, and other specified conditions. This higher incidence highlights the additional physical challenges and complexities that might accompany autism (Kinnear et al., 2020).

The engagement with art of the population, as reported in the study, provides insights into their frequency of visiting museums and galleries, their digital consumption of art, their sensory reactions to art, and the perceived influence of their neurodivergence on their art experiences. First, the autistic participants reported a healthy engagement with museums and art galleries. None stated that they never visit, while 30.43% visit rarely (once a year or less), and an equal percentage visit occasionally (2–4 times a year). Notably, 39.13% of the autistic respondents visit frequently (5 or more times a year), indicating a significant portion of the group actively seeks out physical spaces of art, potentially valuing the sensory and aesthetic experiences offered by such environments.

When it comes to digital engagement with art, 52.17% of autistic individuals view art daily on desktops, tablets, or mobile devices, suggesting a strong preference for integrating art into their daily digital routines. Other engagement levels include 17.39% viewing art 2–3 times a week and another 17.39% 4–6 times a week, with only a small minority (4.35%) never engaging digitally. This pattern indicates a significant inclination toward digital platforms as a medium for art consumption among the autistic population, mirroring the general trend observed in the neurotypical group. In terms of sensory reactions to art, nearly half of the autistic respondents (47.83%) sometimes experience strong sensory reactions such as goosebumps, tingling, or emotional overwhelm when engaging with art. A notable 21.74% feel these reactions most of the time, and 17.39% always have them, suggesting a heightened sensory and emotional engagement with art among many individuals with autism. A smaller portion (13.04%) reported experiencing these strong sensory reactions about half of the time. However, a significant majority of the autistic participants (82.61%) believe their neurodivergence influences the way they perceive or engage with art positively. This overwhelming response highlights the distinctive sensory and cognitive experiences that autism brings to art perception and appreciation, potentially leading to unique insights and connections with the artwork. The remaining 17.39% were unsure of the influence, indicating variability in self-awareness or the impact of neurodivergence on art experiences.

In the group with autism, the reporting of synesthesia is notably higher than in the neurotypical group. Many participants describe experiencing these unusual associations between different senses, with a substantial portion always (21.74%), often (21.74%), or sometimes (21.74%) experiencing synesthetic sensations. This high incidence of sensory crossovers suggests a rich and complex sensory world for many individuals with autism, potentially influencing their perception and appreciation of art. The types of synesthetic experiences reported by autistic individuals are diverse and illustrative of the broad spectrum of synesthesia. Some participants (8.33%) experience specific colors when hearing certain sounds, a phenomenon that vividly merges auditory and visual senses. Another small portion (4.17%) taste flavors in response to specific words or numbers, adding a gustatory dimension to linguistic or numerical stimuli. Feeling tactile sensations, like touch or texture, when observing visual patterns is experienced by 12.50% of participants, indicating a cross-modal interaction between tactile and visual inputs. While some perceive days of the week or months as distinct colors, others associate letters or numbers with particular personalities or emotions, or visualize time units as spatial patterns or objects, each reported by 4.17% to 8.33% of individuals. Though less common, some participants even feel specific textures in response to particular scents. A significant 29.17% of individuals reported other unique synesthetic experiences, highlighting the highly individualistic and varied nature of synesthesia within the autistic community. These experiences range from associating sounds with physical sensations to creating vivid visual imagery in response to auditory stimuli.

One individual vividly describes, "My body moves to sound on its own. I feel every sound as physical sensations. Positive or negative," illustrating a deep, involuntary connection between sound and physical response. Another notes, "I am drawn to certain textures and have a highly visual memory - I have only recently started to link these with some form of synesthesia," indicating a growing awareness of their unique sensory interactions. The range of experiences is profound: one person reports, "In particular, I hear pain," suggesting an auditory experience that manifests as physical pain, while another associates "certain smells to particular images (not ones associated, like, I don't think I smell dog poo if I see it. More like a memory I associate with it, I guess?" This indicates a nuanced interplay between smell and visual or mnemonic associations. Another describes, "Seeing pictures with certain emotions or sounds," combining visual art with an emotional and auditory landscape. Synesthetic experiences like "Taste and color, sound and physical sensation" are mentioned, as well as "Associating soundscapes and curating sound with visuals," highlighting how sensory experiences can deeply influence artistic expression and perception.

These experiences are not only varied but deeply personal, affecting the daily lives and perceptions of the world in significant ways. Participants with autism also reported a range of other unique synesthetic experiences, such as feeling every sound as physical sensations, associating smells with particular images linked to memory, and experiencing colors in relation to taste and sound. Some describe the way these experiences inform their creative expression, curating soundscapes and visuals based on their synesthetic perceptions. These personal accounts highlight the profound effect synesthesia can have on perception and the creative process. In terms of associations, a portion of the autistic group associated numbers or letters with specific colors, tastes, or smells, often with a higher frequency than the neurotypical group. Some participants describe words or phrases evoking particular textures or temperatures and visualizing time as a spatial dimension. These experiences suggest an enhanced and nuanced sensory world where everyday language, numbers, and concepts are intertwined with additional sensory layers.

The impact of synesthesia on the daily lives of individuals with autism varies, with some reporting enhanced memory and recall, more vivid and immersive everyday experiences, and facilitation of creative thinking. However, some face challenges such as confusion or overstimulation in certain environments, leading to avoidance of specific stimuli. A significant number of participants find that synesthesia enhances their appreciation or understanding of music, art, or other sensory experiences, suggesting a positive influence on their aesthetic engagement. In terms of art preferences, the autistic group is drawn to various types of art, with abstract, realistic, mixed media, digital, and tactile art all represented. This diversity reflects the rich aesthetic tastes and sensory preferences within the autistic community. Lastly, emotional reactions to art among autistic individuals tend to be more intense or vivid compared to others, with a large number reporting significantly more intense emotional responses. This heightened emotional engagement with art reflects the deep and often complex connection individuals with autism have with the aesthetic world.

4.3.1.3 ADHD Population

The ADHD population within the study showcases a notable prevalence of co-occurring conditions, reflecting the complex and often intertwined nature of neurodivergent experiences. Dyslexia is reported by 5.45% of individuals with ADHD, indicating challenges with reading that may accompany their attentional characteristics. Dysgraphia, a condition affecting writing abilities, is present in 1.82% of the population, adding another layer to the educational and communication challenges they might face. A significant 41.82% of participants identify solely with ADHD, highlighting it as a prevalent neurotype within the study. Moreover, 9.09% of the population reports experiencing PTSD, suggesting a potential interplay between attentional challenges and trauma responses. Autism is also reported by 14.55% of individuals with ADHD, indicating a notable overlap between these two neurodevelopmental conditions. This overlap is further evidenced by 12.73% of participants identifying with autism and 5.45% with Asperger's syndrome, underscoring the diverse spectrum of cognitive and sensory experiences within the ADHD community. Additionally, 7.27% of respondents specify other neurotypes not listed in the survey, reflecting the broad and varied nature of neurological diversity. These co-occurring conditions and self-identifications paint a detailed picture of the ADHD population, emphasizing the need for understanding and accommodating the rich complexity of their experiences.

The demographic data reveals a rich diversity and co-occurrence with other conditions, providing a complex picture of the lived experience of these individuals. The age distribution of respondents is relatively spread out, with the majority falling between 25 and 44 years of age. This age range suggests that ADHD affects a wide range of adults, impacting their day-to-day lives, careers, and interactions. Notably, a smaller proportion of the group is 65 or older, possibly reflecting generational differences in the diagnosis and understanding. In terms of gender identity, the data indicates a higher representation of females (78.26%) within the population compared to males (17.39%). This finding might reflect a growing recognition of ADHD symptoms and their presentations in females, historically underdiagnosed or misdiagnosed (Mowlem et al., 2019). A small percentage of the population identifies as genderqueer/genderfluid, showcasing the diversity within the community.

Most of the population is not Hispanic/Latinx, with a significant representation of White/Caucasian individuals. The sexual orientation of the group varies, with a notable percentage identifying as heterosexual/straight, while others identify as bisexual or pansexual. The diversity in sexual orientation within the community points to the intersectionality of identities and experiences among neurodivergent individuals. The educational attainment of individuals with ADHD in this study is quite high, with many holding bachelor's and master's degrees, indicating a capable and educated population. The primary language spoken at home is predominantly English, with a small fraction speaking other languages, highlighting the predominance of English-speaking participants in the study.

Physical disabilities are also present in the community, with some individuals reporting visual impairments, hearing impairments, and mobility or physical disabilities. This underscores the importance of considering multiple aspects of diversity and inclusion when addressing the needs and experiences of individuals with ADHD. Comparatively, the demographic shares many similarities with the neurotypical population but also exhibits distinct patterns in gender representation, educational background, and co-occurring conditions. These nuances underscore the need for tailored approaches in educational, occupational, and social contexts to support individuals with ADHD effectively.

With regard to engagement patterns with art, a comparison with the neurotypical population provides insights into how neurological differences might influence aesthetic experiences and art consumption habits. When it comes to visiting museums or art galleries, the ADHD population shows a notable frequency, with the majority occasionally to frequently visiting such places. This indicates a strong interest or inclination toward engaging with art in a traditional setting, not surprising for sensory seekers (Redshaw & McCormack, 2022). Compared to neurotypicals, who also showed a significant engagement with physical art spaces, those with ADHD have a slightly higher percentage of frequently visiting, suggesting that art may serve as a stimulating and enriching experience for them. In terms of digital art consumption, individuals with ADHD demonstrate a high level of daily engagement. This suggests that the convenience, variety, or interactive nature of digital platforms may particularly appeal to those with ADHD, who might seek constant stimulation or novelty (Camp et al., 2021). The majority of both populations view art on their devices daily, but the group has a slightly higher percentage, emphasizing the importance of digital art as a regular part of their lives. The respondents also report strong sensory reactions to art, with a significant number experiencing these reactions most of the time or always. This suggests that individuals with ADHD might have heightened sensory sensitivities or a more intense emotional response to aesthetic stimuli compared to neurotypicals (Panagiotidi et al., 2020). It is notable that a large proportion of the group feels that their neurodivergence positively influences their perception and engagement with art, indicating that population might contribute to a unique and potentially more profound art experience.

For individuals with ADHD, the experience of synesthesia reveals a varied landscape of sensory associations. The population reported significant impacts on their daily lives enhancing memory and recall for 21.43% of respondents. While 7.14% note it makes common experiences more vivid and another 10.71% credit it for facilitating creative thinking, an equal 7.14% face challenges differentiating between sensory inputs. Some individuals, 14.29%, find it creates confusion or overstimulation in certain environments, but an almost equal number, 17.86%, enjoy an enhanced appreciation of music, art, or other sensory experiences. In terms of art preferences, a notable 25.00% are drawn to abstract art, with digital and tactile art each attracting 17.86% of respondents. Meanwhile, emotional reactions to art are reported to be significantly more intense for 39.13% of individuals, indicating a heightened emotional response to artistic stimuli compared to other populations.

Regarding specific experiences, about 8.33% report seeing specific colors when hearing certain sounds, and another 4.17% taste flavors in response to specific words or numbers, suggesting a strong synesthetic experience. Feeling tactile sensations when observing visual patterns is noted by 12.50%, while perceiving days of the week or months as distinct colors is mentioned by 8.33%. Some, about 4.17%, associate letters or numbers with particular personality types or emotions, or visualize time as spatial patterns, also at 4.17%. There is a notable portion, 29.17%, who report other unique synesthetic experiences.

Considering the frequency of the population experiencing numbers or letters with specific colors, tastes, or smells presents a diverse range of experiences, some individuals, 8.70%, always make these associations, while 17.39% often do, and a further 4.35% sometimes find themselves linking these aspects. However, 17.39% rarely make such associations, and a majority of 52.17% never do. Similarly, when it comes to words or phrases evoking particular textures or temperatures, 8.70% always experience this, and 26.09% often feel these unique sensory connections. Only a few, 4.35%, sometimes have these reactions, while 8.70% rarely and 52.17% never experience them. Additionally, 26.09% of individuals always perceive time as a spatial dimension, with 17.39% often and another 17.39% sometimes visualizing time in this manner, indicating a significant portion of the ADHD community experiences time uniquely. The variety in sensory crossover experiences among the population highlights the individualized nature of synesthesia, with each person experiencing and interpreting their sensory world in unique ways.

4.3.1.4 Other Population

The individuals who self-identified with "Other" neurodivergent conditions demonstrate an awareness of neurodiversity without necessarily having an official diagnosis (Fellowes, 2023). These individuals, while not varying significantly from the neurotypical population in demographic data, indicated co-occurrences of conditions such as ADHD, PTSD, autism, and Asperger's syndrome. The group displayed a demographic profile with similarities compared to the neurotypical group. In age distribution, this group had a larger percentage in the 35–44 and 55–64 age ranges compared to neurotypicals, indicating a spread across various adult age groups. The gender identity profile was heavily skewed toward female respondents, similar to the neurotypical group.

When considering racial and ethnic heritage, a vast majority identified as White/Caucasian, akin to the neurotypical group, with small percentages in other racial categories. Sexual orientation among the group showed a high percentage of heterosexual/straight individuals, again related to the neurotypical group, but also included representations of bisexual and pansexual individuals. The highest level of education leaned toward higher degrees, with a significant percentage holding master's degrees, indicating a well-educated cohort. In terms of primary language, English dominated, but a notable percentage spoke other languages at home, suggesting a degree of linguistic diversity. Finally, while most did not report a physical disability, a fraction of the group reported visual impairments or other specified conditions, hinting at the diversity of experiences and challenges within this group.

Regarding their levels of engagement with art, individuals in this group generally share similarities with the neurotypical group, although they exhibit discernible inclinations. When it comes to visiting museums or art galleries, none of the respondents in this group reported never visiting, and a balanced distribution exists between those visiting rarely, occasionally, and frequently. This indicates a generally robust involvement with physical art venues, mirroring the behavior of the neurotypical group, albeit with a slightly elevated occurrence of occasional and regular visits. Regarding digital art consumption, this group also integrates art into their daily lives, with a substantial percentage viewing art on devices daily and others engaging multiple times a week. This indicates a significant presence of art in their everyday activities and mirrors the high level of digital art engagement seen in the neurotypical group. In terms of experiencing strong sensory reactions to art, a majority reported sometimes or more frequent strong reactions, including goosebumps, tingling, or emotional overwhelm, similar to the neurotypical group. However, this group has a slightly higher proportion of individuals who always or most of the time experience such strong sensory reactions when engaging with art, suggesting a potentially heightened sensitivity or more intense interaction with aesthetic experiences. Importantly, when considering the influence of their self-identified neurodivergent condition on their perception or engagement with art, a majority believe it positively influences their experience.

The group reported various sensory crossover experiences. A fraction of the group, 7.69%, always experience sensory crossovers or synesthesia, with an additional 23.08% often experiencing these unique associations between different senses. This indicates that a significant portion of this group is regularly navigating a world enriched with these extra layers of sensory interaction, compared to the neurotypical population. When detailing the types of synesthetic experiences, 16.67% reported seeing specific colors when hearing sounds, and another 16.67% perceived days of the week or months as distinct colors. Additionally, 11.11% felt tactile sensations when observing visual patterns. These responses highlight the rich and varied sensory world some individuals navigate, with specific sounds, days, or patterns evoking vivid colors or tactile sensations. In terms of how synesthesia affects their daily life or perception of the world, 21.05% reported that it enhances memory and recall capabilities, while 15.79% find that it facilitates creative thinking or artistic endeavors. However, 10.53% noted that it sometimes results in discomfort or sensory overload, indicating that while synesthesia can enrich experience, it can also present challenges.

Regarding art preferences, 26.32% are drawn to digital art, followed by 21.05% preferring realistic art. Emotional reactions to art also appear more intense for this group, with 38.46% reporting significantly more intense reactions compared to the experiences of other populations. This heightened emotional response underscores the deep impact that art can have on individuals with diverse neurological backgrounds. Overall, the "Other" group experiences a vivid and sometimes intense sensory world, often enhanced by synesthetic perceptions. Their engagement with art is characterized by a strong emotional component and a tendency to experience art in a uniquely personal way, shaped by their individual sensory and emotional landscape.

4.3.1.5 Post-Traumatic Stress Disorder Population

The group identifying with PTSD reflects a demographic slightly older than the neurotypical population, with the majority being between 45 and 54 years old. There is a significant representation of females, accounting for 81.82% of the respondents, which aligns with general trends, where the condition is often seen at a higher rate in women (Hiscox et al.,

2023; Hu et al., 2023; Johnson et al., 2023). Additionally, most of the respondents were not Hispanic/Latinx (90.91%) and predominantly identified as White/Caucasian (83.33%), with a small percentage indicating other European or mixed ethnic backgrounds. In terms of sexual orientation, the majority identified as heterosexual/straight (63.64%), with some identifying as queer or bisexual, which underscores the diversity within this group. Their educational background is notably varied, with a good number having a master's degree (36.36%) or a bachelor's degree (27.27%). This suggests a well-educated cohort, yet it is worth noting the presence of individuals with doctoral degrees (18.18%) as well as those with some college but no degree (18.18%).

Language-wise, English dominates as the primary language spoken at home, which is consistent across most survey groups. Interestingly, a small portion of the group speaks multiple languages at home, indicating a diverse linguistic background. When considering physical disabilities, a notable portion of this group reported having one, with mobility or physical disability, visual impairment, and other specified conditions like lower back issues, vertigo, and erythromelalgia mentioned.

Individuals with PTSD showed varying degrees of engagement with art, both physically in museums or art galleries and digitally via devices. About a third of the participants reported rarely visiting physical art spaces, while another third occasionally visit 2–4 times a year, and the remaining frequently engage five or more times a year. This indicates a wide range of engagement levels, with a notable portion integrating art visits into their regular activities. Digital art consumption was more prevalent among the group, with the majority viewing art on desktops, tablets, or mobile devices daily or several times a week. This suggests that for some, digital platforms may provide a more accessible or comfortable means of engaging with art.

In terms of sensory reactions to art, the group reported strong responses, including goosebumps, tingling, or emotional overwhelm, with a significant percentage experiencing these reactions sometimes to always. This could imply a heightened sensory awareness or emotional connection to art among individuals with PTSD, possibly influenced by their neurodivergence. Most respondents believe their neurodivergence positively influences their perception or engagement with art, indicating that their unique experiences may enrich their interaction with artistic expressions. However, a small percentage felt negatively impacted or were unsure about the influence, reflecting the varied ways PTSD can affect art experience.

Turning to reports of those with PTSD and reported synesthetic experiences, more than half of the participants have experienced these sensory "crossovers" at least sometimes, with a significant portion often or always experiencing such phenomena. Commonly reported sensations included seeing specific colors when hearing certain sounds and feeling specific textures when exposed to particular scents.

One notable aspect of the reported synesthesia in this group is the impact on daily life and perception of the world. Many indicated that these experiences enhance memory and recall capabilities or make every day experiences more vivid and immersive. A good portion of the group also mentioned that synesthesia enhances their appreciation or understanding of music, art, or other sensory experiences, though some also reported challenges in differentiating between sensory inputs.

In terms of art preferences, abstract art was popular among the group, followed by other styles like impressionism and realism. A majority of the group found that their emotional reactions to art were significantly more intense or vivid compared to others, suggesting a deep connection and response to artistic stimuli. These insights into the sensory and emotional experiences of individuals with PTSD concerning synesthesia and art engagement highlight the complexity of sensory processing in this group and underscore the need for nuanced understanding and accommodations in artistic and therapeutic contexts.

4.3.1.6 Dyslexic Population

The demographic data of individuals with Dyslexia in the study provides a snapshot of this specific neurotribe. The age distribution of respondents varied, with the largest groups falling between 35–44 and 45–54 years. There was a higher percentage of females (66.67%) compared to males (33.33%), and all respondents identified as heterosexual, bisexual, or asexual. A small percentage identified as Hispanic/Latinx, with the majority being non-Hispanic/Latinx. Racially, the majority of participants identified as White/Caucasian with a small portion as American Indian or Alaskan Native or other specified ethnicities. All respondents indicated English as the primary language spoken at home, showcasing a lack of linguistic diversity within this small sample. Educational backgrounds varied, with some individuals holding high school diplomas, associate degrees, bachelor's degrees, master's degrees, and even doctoral degrees.

Regarding physical disabilities, half of the individuals reported none, while others mentioned having visual or hearing impairments or mobility or physical disabilities. This portrays a scenario of a moderately diverse group with variations in age, education, and physical abilities, though exhibiting less diversity in terms of ethnicity, language, and gender.

Individuals with Dyslexia in the study shared their experiences with art, providing insight into how they engage with the visual arts. Half of the respondents reported visiting museums or art galleries rarely, once a year or less, while a third frequently visited more than five times a year, indicating a range of engagement levels. When it comes to digital consumption of art, a significant two-thirds of the individuals viewed art daily on desktops, tablets, or mobile devices, suggesting a high level of integration of art into their daily digital experiences. In terms of sensory reactions to art, the majority (83.33%) sometimes experienced strong sensory reactions like goosebumps, tingling, or emotional overwhelm when engaging with art, with a smaller portion (16.67%) feeling these reactions most of the time. This indicates that while strong sensory experiences with art are common among this group, they are not universally intense or always occurring.

When asked about the influence of their neurodivergence on their perception or engagement with art, respondents were divided. A third felt their neurodivergence positively influenced their art experiences, highlighting a beneficial connection between their neurodivergence and art perception. An equal number were unsure, and a smaller fraction reported no discernible influence. This suggests a diversity of experiences and perceptions among these individuals regarding how their condition interacts with their appreciation and engagement with art. Some see it as enhancing their experience, others are uncertain, and some do not see a significant impact.

The population also reported varied sensory crossover experiences. About one-sixth of the participants always experience these unique sensory associations, with a third sometimes experiencing them, indicating a notable presence of synesthetic experiences in this group. Specifically, seeing specific colors when hearing certain sounds was the most reported sensation, followed by feeling tactile sensations when observing visual patterns and visualizing time units as spatial patterns or objects. Interestingly, a significant portion of respondents (83.33%) never associated numbers or letters with specific colors, tastes, or smells, suggesting this particular type of synesthetic experience is less common among them. In terms of how synesthesia impacts their daily life, respondents noted that it enhances memory and recall capabilities and provides a unique way to relate to and communicate with others. Some respondents also mentioned that synesthesia makes ordinary experiences more vivid and immersive, while others reported challenges in differentiating between sensory inputs and occasionally resulting in confusion or overstimulation.

When discussing types of art drew their attention, digital media was a popular choice, followed by tactile art forms, showing a diverse interest in various art mediums. Among this group, emotional responses to art were notably intense for a subset, with one-third of the respondents reporting significantly stronger reactions than their peers. However, a portion of the group expressed uncertainty regarding the intensity of their emotional responses. This implies a spectrum of emotional involvement with art within the Dyslexic community, ranging from highly intense to uncertain.

4.3.2 Synesthetic Conditions

4.3.2.1 Non-Synesthete Population

Building on the section reviewing the association of neurological conditions with various aesthetic experiences, respondents in the following were sorted into various categories based on their synesthetic conditions and whether they reported experiencing crossover sensations or synesthesia. This section opens by focusing on the demographics and experiences of individuals who reported not having synesthetic symptoms, serving as a baseline for understanding the prevalence and nature of synesthetic experiences across different neurological conditions. Demographically, nearly a third of all respondents (32%) indicated they do not experience synesthesia, suggesting a significant portion of the population does not associate with these sensory crossover experiences. The largest portion of these respondents identified as neurotypical, but it is noteworthy that the absence of reported synesthetic experiences did not preclude diagnoses of other neurological conditions. These diagnoses included dyslexia, ADHD, PTSD, and autism. The age distribution of these non-synesthetes ranged from 18 to over 65, with a concentration in the 25–44 age group, reflecting the broader adult population's distribution. Most identified as female (70%), and a vast majority were not Hispanic/Latinx (94.74%) and identified as White/Caucasian (95.24%). The sexual orientation of the majority was heterosexual/straight (80%), and the respondents were predominantly English speakers (95%). In terms of education, the group was well-educated, with a significant number holding bachelor's (40%) or higher degrees.

Individuals who do not possess synesthetic experiences, often referred to as non-synesthetes, cultivate a distinct connection with art that sets them apart from those who encounter sensory crossovers. According to the study, a small portion (5%) of these respondents never visit museums or art galleries, while a more considerable number engage with art to varying degrees: 30% rarely, another 30% occasionally, and 35% frequently visit such cultural venues. This suggests a substantial engagement with physical art spaces among non-synesthetes. When it comes to digital consumption, none reported never viewing art on desktops, tablets, or mobile devices for enjoyment, with 50% engaging daily, indicating a significant tilt toward the regular integration of digital art into their lives. This preference for digital media might reflect the convenience or accessibility of experiencing art in this format. On the other hand, despite lacking synesthetic experiences, 70% sometimes and 25% most of the time reported having strong sensory reactions to art, like goosebumps or emotional overwhelm, indicating that while they do not experience sensory crossovers, art profoundly affects them. Only a minority (23.53%) felt that their neurodivergence positively influenced their perception or engagement with art, with an equal percentage unsure of its impact and a significant proportion (41.18%) finding the question not applicable, potentially indicating a self-identification as neurotypical within this group.

The data portrays non-synesthetes as actively engaging with both physical and digital art forms, experiencing strong, albeit not synesthetic, reactions to art, and holding varied views on how their neurodivergence might influence their artistic experiences. Their engagement patterns highlight the universal appeal of art across different sensory experiences and neurological backgrounds. Overall, the baseline demographic establishes a comparative group for analyzing how synesthetic experiences vary among different neurological conditions and provides a foundation for understanding the broad spectrum of sensory experiences across the population. The following sections will delve into the synesthetic experiences reported by individuals with various neurological conditions, from the most to the least common, providing insights into the intricate relationship between neurological diversity and sensory perception.

4.3.2.2 Chromosthesic Synesthetes

Chromesthetics, individuals who experience synesthesia particularly related to color and sound, represent a distinct and significant portion of the synesthetic population, second only to non-synesthetes in this study. Examining the demographic makeup of chromesthetics provides an insight into their general profile. Predominantly female (72.73%) with a majority falling within the 35–44 and 55–64 age ranges, these individuals predominantly identify as heterosexual (81.82%) with a few identifying as bisexual or pansexual. Most chromesthetics (90.91%) are not Hispanic or Latinx, with a significant majority (58.33%) identifying as White/Caucasian, along with diverse minority representations. Educationally, they tend to be highly educated, with over half (54.55%) holding a master's degree and a notable percentage possessing a bachelor's degree. English is the primary language for most (72.73%), with a minority communicating in other languages at home, indicating a degree of linguistic diversity within this group. When it comes to physical disabilities, the vast majority (80%) do not report having any, suggesting that their synesthetic experiences are more linked to neurological rather than physical conditions. This demographic snapshot reveals a group that is diverse yet shares common threads in higher education levels and a predominant gender bias toward females. Such details provide a backdrop for understanding how chromesthesia influences and is influenced by various aspects of their lives, including their interaction with art and sensory experiences.

For those reporting experiences of chromesthesia, synesthesia intertwines deeply with daily life and shapes their perception of the world in various nuanced ways. A notable 22.73% of respondents reported that synesthesia enhances their appreciation or understanding of music, art, or other sensory experiences, turning everyday life into a more enriched experience. For these individuals, colors might dance to melodies, or landscapes of hue unfold with every note, deepening their engagement with creative arts.

Facilitating creative thinking or artistic endeavors was another significant aspect reported by 18.18% of the chromesthetics. Synesthesia often provides a unique sensory palette that can inspire innovation and creativity, leading to a distinct approach in artistic expression and problem-solving.

However, synesthesia is not without its challenges. While 9.09% acknowledge that it enhances memory and recall capabilities, likely through vivid associative colors or patterns for letters, numbers, or sounds, an equal percentage sometimes find it leading to discomfort or sensory overload. This indicates that while synesthesia can be a cognitive advantage, it can also lead to overwhelming sensory experiences in certain environments (Rinaldi, 2021). Additionally, 13.64% of chromesthetics find that synesthesia provides a unique way to relate to and communicate with others, possibly allowing for more empathetic and enriched interpersonal connections. On the flip side, a small portion (4.55%) reports that it leads to avoidance of specific stimuli or environments, possibly due to unpleasant or overwhelming sensory associations.

Chromesthesia not only exerts an impact on everyday life but also distinctly shapes preferences in art and amplifies the intensity of their emotional responses. When it comes to art preferences, chromesthetics display a varied taste. A notable 23.08% each are drawn to realistic art and digital media (photography, film, AI-generated), indicating a spread across traditional and modern mediums. Abstract and mixed media also appeal to 15.38% each, possibly due to their vibrant use of colors and forms that may resonate with the vivid sensory experiences of chromesthesia. A smaller portion, 7.69%, find themselves drawn to tactile art, suggesting an appreciation for the sensory depth and texture. Interestingly, one respondent noted a particular draw toward music, underscoring chromesthesia's often close link to auditory stimuli and the vivid, color-laden experiences it can invoke. This aligns with the broader theme of chromesthetics seeking art forms that stimulate their unique sensory interplay.

Emotionally, the impact of chromesthesia on art perception is pronounced. A significant 27.27% report that their emotional reactions to art are significantly more intense compared to others, with another 45.45% noting somewhat more intense feelings. This majority indicates that chromesthesia can amplify the emotional resonance of art, perhaps due to the additional sensory dimensions it adds to the experience. Only 9.09% feel their emotional reactions are comparable to others, suggesting that for most chromesthetics, the condition adds a layer of depth and intensity to their art experiences. However, despite the heightened emotional engagement, there's an element of uncertainty for 18.18% of respondents, reflecting the varied and personal nature of synesthetic experiences. The uncertainty might stem from difficulties in articulating or comparing their unique perceptions with those of non-synesthetes.

4.3.2.3 Mirror-Touch Synesthetes

Mirror-touch synesthetes, who experience tactile sensations like touch or texture when observing visual patterns, are an intriguing group with their own demographic spread. Among them, ADHD (30%), ASD (10%), and other unspecified neurodivergent conditions (20%) are present, suggesting a correlation between these conditions and synesthesia. Interestingly, a significant portion (30%) identifies as neurotypical, highlighting the broad range of people who experience synesthetic sensations. The age distribution of mirror-touch synesthetes leans toward the middle-aged, with the majority falling in the 25–34 and 55–64 age brackets, making up 37.50% and 25.00% respectively. Gender distribution shows a majority of females (62.50%) and a smaller representation of males (37.50%). Almost all individuals (85.71%) are not Hispanic or Latinx, and a unanimous 100% identify as White/Caucasian, indicating a potential need for more diverse research in this area. In terms of sexual orientation, the majority are heterosexual/straight (66.67%), with a smaller representation of bisexual (22.22%) and queer (11.11%) individuals. This suggests that while mirror-touch synesthesia is experienced across a variety of sexual orientations, it may be more prevalent or recognized in certain groups.

Educationally, most mirror-touch synesthetes have pursued higher education, with bachelor's and master's degrees each accounting for 37.50%. This may indicate a higher level of awareness or self-diagnosis in individuals who have had more educational opportunities. Most speak English at home (75%), with a smaller portion speaking Portuguese (25%), suggesting a primary Western cultural background. Physically, the vast majority (87.50%) report no disabilities, with a small percentage (12.50%) reporting visual impairment. This demographic snapshot provides a foundation for understanding the typical profile of a person with mirror-touch synesthesia, offering insights into the variety of backgrounds from which these individuals come.

The impact of synesthesia on daily life and perception of the world is as varied as the condition itself, with individuals reporting a range of effects from highly beneficial to challenging. A significant portion (17.65% each) of synesthesia elevates memory and recall capacities, infusing ordinary experiences with heightened vividness and immersion. This can turn routine tasks into sensory-rich experiences and potentially aid in learning and creativity. At the same time, crossover experiences are not without their challenges. Some individuals (5.88%) find it poses difficulties in differentiating between sensory inputs, which can be disorienting or overwhelming. Additionally, about 11.76% of individuals report that synesthesia leads to confusion or overstimulation in certain environments, indicating that highly sensory environments can be particularly challenging. Notably, synesthesia seems to nurture a distinctive bond with the world for certain individuals. Around 11.76% of respondents feel it allows them to relate and communicate with others in unique ways, possibly through shared sensory experiences or more profound empathy. For those creatively inclined, it can facilitate artistic endeavors (5.88%), enhancing the creative process through a richer sensory palette.

Avoidance behaviors are noted by 5.88% of individuals who might steer clear of specific stimuli or environments that trigger overwhelming synesthetic responses. Similarly, another 11.76% sometimes experience discomfort or sensory overload, indicating that while synesthesia can be enriching, it can also be demanding and requires careful navigation of the sensory world. A small proportion (5.88%) feel that synesthesia enhances their appreciation or understanding of music, art, or other sensory experiences, suggesting that for some, this condition can deepen the aesthetic experience. Conversely, an equal percentage see no significant influence or consider their synesthesia mostly neutral, indicating that the impact of synesthesia can range from transformative to negligible.

These synesthetes also demonstrate unique art preferences and emotional reactions, often influenced by their rich sensory experiences. A striking 44.44% are drawn to digital art, including photography, film, and AI-generated pieces, possibly because of the vivid and dynamic nature of these mediums that might resonate with their synesthetic experiences. Additionally, 22.22% enjoy mixed media, and another 22.22% are attracted to tactile art forms, suggesting a preference for art that engages multiple senses or has a pronounced physicality.

In terms of emotional reactions to art, many mirror-touch synesthetes report intensified responses. Half of the individuals (50.00%) feel somewhat more intense emotions compared to others, with 12.50% describing their reactions as significantly more intense. This heightened emotional resonance with art could be attributed to their synesthetic experiences, which blend sensory perceptions in everyday life, making them perhaps more attuned to the sensory and emotional layers within art (Banissy et al., 2011). A quarter (25.00%) of mirror-touch synesthetes find their emotional reactions to art comparable to others, indicating a diversity in how synesthesia affects emotional engagement with art. Additionally, a small portion (12.50%) are unsure about the intensity of their emotional reactions, reflecting the subjective and personal nature of both synesthesia and emotional response to art.

One quote encapsulates the enduring influence of their neurological uniqueness on their appreciation of art: "I believe my ADHD/neurodiversity is the very reason I fell in love with music. While I have an appreciation for various music genres, my strongest connection lies with electronic music due to its intricate layers. I often describe it as a gentle tickle to my brain. Curiously, I often find that I require background music to concentrate because it engages the part of my brain that typically seeks distractions.“ This reflection not only emphasizes the profound bond with specific art forms but also showcases how synesthetic and neurodivergent experiences can enhance and deepen one's relationship with art, turning it into a source of fascination, emotional resonance, and even heightened cognitive focus.

4.3.2.4 Grapheme-Color Synesthetes

Grapheme-color synesthetes, who perceive days of the week or months as distinct colors (Fig. 4.2), present an interesting demographic profile. The group consists entirely of females, with the majority being Caucasian (83.33%) and heterosexual (83.33%). Most of the individuals are non-Hispanic/Latinx (100%), primarily English speakers at home. Their ages range across the spectrum, with a notable presence in the 55–64 age group (33.33%). In terms of education, the group is diverse, with some having no high school diploma (16.67%) and others holding doctoral degrees (16.67%). Both bachelor's and master's degrees figure prominently in their educational background. This indicates a variety of life experiences and knowledge bases within the grapheme-color synesthetic community. A significant 83.33% report no physical disabilities, which might suggest that their synesthetic experiences are predominantly cognitive and sensory, rather than linked to any physical conditions. However, one individual does mention having Interstitial Cystitis, highlighting that even within specific neurological experiences like synesthesia, there can be a range of other health conditions present.

For these individuals, their condition significantly influences their daily life and perception of the world. The most prominent effect reported by 40% of respondents is enhanced memory and recall capabilities. This likely stems from their ability to associate colors with days of the week

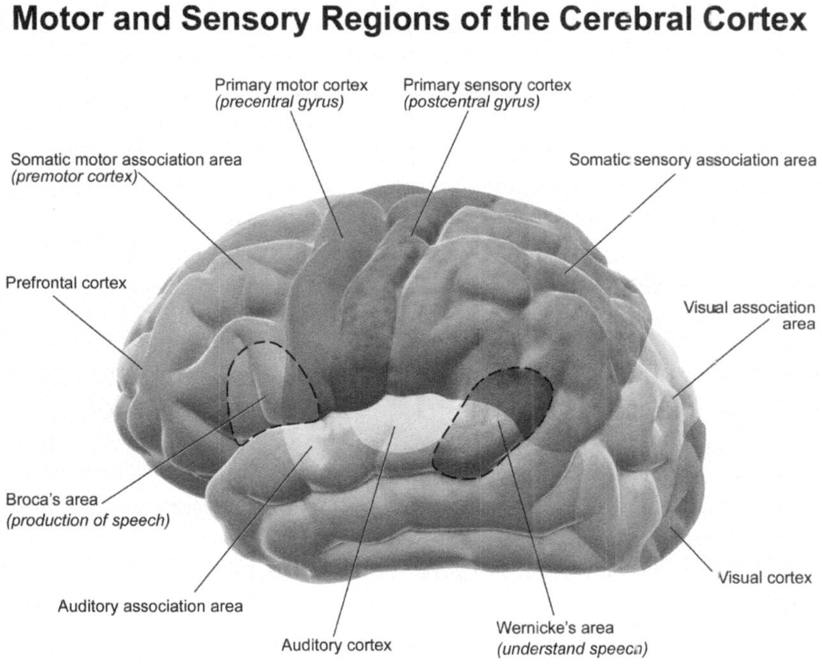

Fig. 4.2 Synesthesia months of the year (*Source* From Wikimedia Commons, licensed under CC-BY 2)

or months, providing them with a vivid and organized mental calendar. About 10% find that synesthesia makes ordinary experiences more vivid and immersive. This might be because ordinary moments are infused with an additional layer of sensory input, turning routine into something more colorful and engaging. Another 10% acknowledge that their condition facilitates creative thinking or artistic endeavors. The unique sensory experiences might foster a deeper connection to creative work, enabling them to think and create in ways that others might not.

However, the condition presents its share of challenges. While none of the respondents reported difficulties in discerning between sensory inputs specifically, 10% do encounter moments of confusion or overstimulation in certain environments. This could be attributed to the overwhelming nature of constantly experiencing color associations. Similarly, 10% occasionally find themselves in a state of discomfort or sensory overload, suggesting that while synesthesia can enhance experiences, it can also intensify them to uncomfortable levels. Interestingly, 10% offer a unique perspective, indicating that synesthesia provides a means to connect with and communicate with others. This connection might be forged through shared experiences with fellow synesthetes or by contributing a distinctive viewpoint to conversations. Another 10% perceive their synesthesia as having a limited impact or being mostly neutral in their lives, underscoring a spectrum of influence that ranges from highly significant to relatively inconsequential.

Moreover, the population shows varied preferences in art forms and report heightened emotional responses compared to others. When it comes to types of art, abstract and mixed media, each attracts 28.57% of individuals, perhaps due to the abstract nature of their own sensory experiences aligning with the less structured, more interpretive nature of these art forms. Digital art is equally preferred, also engaging 28.57% of respondents. This might suggest an attraction to the vibrant colors and dynamic visuals often found in digital works. Realistic art attracts a smaller portion, about 14.29%, indicating a lesser preference for traditional representations of reality.

The impact of the condition on emotional reactions to art is notably profound. A significant 66.67% of individuals experience somewhat more intense emotional reactions to art compared to the general population, while 16.67% report significantly more intense reactions. This heightened emotional response might be due to the additional layer of sensory experience that synesthesia brings to viewing art, making the experience more immersive and personally resonant. Only a small portion, 16.67%, feel their emotional reactions are comparable to others, suggesting that for most Grapheme-Color synesthetes, art isn't just seen; it is deeply felt in a more vivid and intense manner. This data reflects the enriched sensory world of grapheme-color synesthetes, where art becomes a more profound experience, filled with intense emotional and sensory engagement. Their unique perception of the world adds depth to their appreciation and understanding of art, making every piece more than just a visual experience.

4.3.2.5 Lexical-Gustatory Synesthetes

Lexical-gustatory synesthetes, those who taste flavors in response to specific words or numbers or feel certain textures when exposed to particular smells are a unique group with diverse backgrounds. The group comprises individuals with various neurodivergent conditions, including ADHD and autism, each accounting for 20% of the group. PTSD and individuals identifying with autism also represent significant portions of this group. Demographically, the group is primarily female, making up 66.67% of respondents, with a smaller male representation at 16.67%. The age distribution is quite spread out, with a notable concentration in the 45–54 and 55–64 age brackets, each accounting for 33.33%. The group is primarily non-Hispanic/Latinx (100%) and predominantly identifies as White/Caucasian (37.5%), followed by Asian (25%). In terms of sexual orientation, there's a split with 50% identifying as heterosexual/straight and a significant portion (33.33%) identifying as pansexual. The highest level of education achieved by most is doctoral, representing 50% of the group, followed by those with master's degrees (33.33%). This indicates a highly educated demographic. The majority speak English at home, with one individual specifying Italian as their primary language. Most do not have a physical disability, although a small percentage reported co-occurring conditions such as erythromelalgia.

These synesthetes encounter a variety of effects stemming from their synesthesia that influence their daily lives and how they perceive the world. A portion of them, accounting for 15.38% of responses each, report that their synesthesia enhances their memory and recall abilities, as well as intensifies the vividness and immersiveness of their everyday experiences. This indicates that, for some individuals, synesthesia enriches their existence by imbuing routine activities with additional layers of sensory involvement, thereby turning them into more vibrant and unforgettable occasions. For others, synesthesia facilitates creative thinking and artistic endeavors, also reported by 15.38% of individuals. This aspect of synesthesia often leads to innovative problem-solving techniques and unique artistic expression, leveraging the ability to blend senses in unconventional ways. However, it is not without challenges. Some individuals, each representing 7.69%, report that it poses challenges in differentiating between sensory inputs, creates confusion or overstimulation in certain environments, leads to avoidance of specific stimuli or environments, or results in discomfort or sensory overload. These aspects highlight the complexity of living with synesthesia, where the blending of senses can sometimes be overwhelming or disorienting. A small portion believes synesthesia enhances their appreciation or understanding of music, art, or other sensory experiences, suggesting that it can deepen emotional and aesthetic responses to various forms of art. Others indicate the condition poses no significant influence or is mostly neutral to their daily lives. The diversity in responses underscores the highly individual nature of synesthesia and how its impact can vary widely among different people.

Lexical-gustatory synesthetes exhibit specific inclinations toward various art forms, reflecting how their unique sensory experiences shape their aesthetic preferences. A significant 42.86% find themselves drawn to abstract art, possibly because the non-representational nature of abstract art allows for a freer association of sensory experiences and interpretations. This is followed by 28.57% who prefer realistic art, suggesting a connection to more literal representations that might resonate with their sensory experiences. Mixed media and tactile art each attracts 14.29%, indicating an appreciation for art forms that engage multiple senses or offer a more physical, texture-rich experience.

When it comes to emotional reactions to art, the majority of these synesthetes, 66.67%, report significantly more intense responses compared to others. The heightened sensitivity underscores the profound impact art can have on individuals with synesthesia, possibly due to their more complex sensory processing. A smaller portion, 16.67%, feel their emotional reactions are somewhat more intense, while another 16.67% feel their reactions are comparable to others. This variability reflects the personal nature of both art appreciation and synesthetic experiences. Whether drawn to the ambiguity of abstract art or the tangible textures of tactile pieces, their preferences highlight the deep connection between synesthesia and the emotional and sensory dimensions of art appreciation.

4.3.2.6 Ordinal-Linguistic Synesthetes

Ordinal-linguistic synesthetes, individuals who associate letters or numbers with particular personality types or emotions, are a unique demographic in the study of synesthesia. The small sample of individuals identifying with this type of synesthesia is predominantly female (100%), with no reported male or non-binary individuals. In terms of age, the group shows a spread across different age brackets, with a notable absence of individuals aged 45–54 and 65 or older. All respondents identified as non-Hispanic/Latinx and predominantly spoke English at home, highlighting a certain uniformity in cultural and linguistic background. Their education levels varied, with a concentration in higher education; 50% hold a bachelor's degree, and another 25% have a master's degree, indicating a relatively well-educated group. The sexual orientation of the respondents is diverse, with 50% identifying as heterosexual/straight, and the rest divided between bisexual and pansexual. Uniquely, all respondents do not report having any physical disabilities, suggesting that their synesthetic experiences are primarily cognitive or sensory rather than linked to physical conditions. The demographic makeup of the population reveals a group that is predominantly female and educated, with a diverse range of ages and sexual orientations, predominantly speaking English and without physical disabilities. The profile provides a glimpse into the varied backgrounds of individuals who experience the world through the unique lens of associating abstract concepts like letters and numbers with emotions and personalities.

These synesthetes experience the world in a uniquely multidimensional way, where letters and numbers evoke specific personalities or emotions. This synesthetic experience influences their daily lives and perception of the world in various ways. The most significant impact, as reported by 26.67% of individuals, is the enhancement of memory and recall capabilities. This suggests that associating characters with emotions or personalities may create more robust mental links, aiding memory retention (Pearson, 2019). Around 13.33% of individuals find that their synesthesia makes every day experiences more vivid and immersive, while an equal percentage feel that it facilitates creative thinking or artistic endeavors. This indicates that for some, synesthesia adds a layer of richness and complexity to daily life and the creative process, offering a deeper connection to the world around them. However, synesthesia is not without its challenges. While no one reported specific issues with sensory input differentiation, about 6.67% experience confusion or overstimulation in certain environments, which could be due to the overwhelming nature of constant sensory cross-connections. Similarly, 13.33% sometimes face discomfort or sensory overload, indicating that the condition can become taxing in specific contexts.

A portion of the group (13.33%) also appreciates the unique social interactions their synesthesia facilitates, as it provides a different perspective and way to relate to others, potentially leading to more profound and engaging conversations. However, 6.67% tend to avoid certain stimuli or environments, suggesting that the sensory experiences triggered by their synesthesia can be unpleasant or distracting in specific contexts. Lastly, a small percentage (6.67%) feel that their synesthesia enhances their appreciation or understanding of music, art, or other sensory experiences. This demonstrates the ability of the condition to enrich the aesthetic and emotional experience of art and the world. Overall, they navigate a world filled with added layers of sensory experiences, which can enrich their lives and creativity but also pose unique challenges. The way they interact with and interpret their surroundings is deeply influenced by the intertwining of senses and cognition, shaping their memory, creativity, and daily experiences.

Ordinal-linguistic synesthetes also display diverse preferences and emotional responses to art. Their unique perception influences the types of art they are drawn to, as well as the intensity of their emotional reactions to artistic works. In terms of art preferences, a significant 50% of the individuals are drawn to mixed media, suggesting an affinity for art forms that blend various elements, perhaps mirroring the blended sensory experiences in their daily lives. Both abstract and digital (photography, film, AI-generated) art appeal to 16.67% of the respondents, indicating a preference for art forms that might offer complex, layered, or unconventional sensory experiences. A similar percentage (16.67%) appreciate realistic art, suggesting that for some, the clear, defined nature of realism resonates with their synesthetic experiences.

The synesthetes in this study experience a diverse array of effects resulting from their neurology, exerting a notable influence on their daily routines and perceptual processes. It is noteworthy that 15.38% of respondents each report that their synesthesia has the effect of augmenting their memory and recall capacities, while concurrently amplifying the intensity and immersive nature of their quotidian experiences. These findings amplify the knowledge that for certain individuals, synesthesia imparts a heightened dimension to their lives by infusing mundane activities with supplementary layers of sensory engagement, consequently rendering them more vivid and indelible occasions.

4.3.2.7 Time–Space Synesthetes

Time–space synesthetes, individuals who visualize time units as spatial patterns or objects, are a unique subset of synesthetes with distinct demographic characteristics. This group comprises individuals diagnosed with ADHD, autism, and other unspecified conditions, as well as those who consider themselves neurotypical. The majority are in their mid-life, split evenly between the 25–34 and 35–44 age groups, and predominantly female. Educationally, these individuals are quite accomplished, with half holding a master's degree and a quarter having a doctoral degree. All of them communicate primarily in English at home, with a few reporting French as another language spoken. The group shows a wide range of sexual orientations, predominantly heterosexual and pansexual, and they come from varied ethnic backgrounds, although the majority are White/Caucasian. In terms of physical disabilities, the majority report none, but a few have visual impairments. This diverse group represents a cross-section of society, yet they share the unique cognitive phenomenon of perceiving time in spatial constructs, which significantly impacts how they interact with and perceive the world around them.

These synesthetes experience the world in a unique way that significantly impacts their daily life and perception. A notable 18.75% of individuals report that their synesthesia enhances memory and recall capabilities, suggesting that visualizing time as spatial constructs aids in information retention and retrieval. Another 12.5% experience a vividness and immersion in their daily experiences, indicating that this form of synesthesia can enrich everyday life with added sensory dimensions. However, this unique perception is not without its challenges. A similar 18.75% find themselves confused or overstimulated by their environment, likely due to the constant influx of sensory information processed in an uncommon manner. This can lead to avoidance of specific stimuli or environments as reported by 12.5%, demonstrating a need to manage overwhelming sensory input.

The condition also fosters a connection with others for 12.5% of individuals, suggesting that sharing and communicating their experiences can be a source of bonding or mutual understanding. Despite the challenges, some find that synesthesia facilitates creative thinking and artistic endeavors, as well as enhances the appreciation of music, art, and other sensory experiences, each reported by 6.25%. On the other hand, a small percentage sometimes experience discomfort or sensory overload, emphasizing the need for coping mechanisms or strategies to navigate their heightened sensory world. Overall, while the impacts of the condition vary widely among individuals, it undeniably shapes their interaction with the world in profound ways.

These synesthetes exhibit diverse art preferences, often seeking out forms that resonate with their unique sensory experiences. A significant 40% of individuals are drawn to mixed media, perhaps due to its complex, layered nature which might parallel the intricate way they perceive time and space. Additionally, abstract and digital art each attracts 20% of individuals, suggesting a preference for genres that challenge traditional perceptions of space and reality, aligning with their synesthetic experiences.

In terms of emotional reactions to art, responses are varied. A quarter of the group reports significantly more intense reactions compared to others, suggesting that their synesthetic experiences might amplify the emotional impact of art. Half of the individuals feel their emotional responses are comparable to others, indicating that while their perception of time and space is unique, it does not necessarily intensify all emotional responses to art. Lastly, a quarter are unsure about the intensity of their reactions, reflecting the personal and subjective nature of both synesthesia and art appreciation. These preferences and emotional experiences of the group with art are as varied as their synesthetic experiences, reflecting a wide spectrum of interactions and responses to different art forms. Whether it be mixed media, abstract, or digital art, these individuals engage with art in ways that might amplify or uniquely interact with their perception of the world.

4.3.3 Survey Study Summary

The exploration of sensory profiles in relation to neurological conditions and synesthesia presents a fascinating intersection of neurology, psychology, and personal experience. This section delves into the diverse ways individuals experience the world around them, influenced by their unique neurological makeups and synesthetic perceptions. By examining the relationship between various neurological conditions, demographic factors, and the multifaceted world of synesthesia, we gain insights into the rich tapestry of human perception, uncovering how these factors contribute to the way art is experienced and appreciated. This introductory exploration sets the stage for a deeper understanding of the complex interplay between the brain, senses, and the external environment.

The research indicates a notable intersection between neurological conditions such as ADHD, ASC, and PTSD with the occurrence of synesthesia. This connection suggests a potentially higher prevalence of synesthetic experiences among individuals with these conditions, highlighting the need for a nuanced understanding of how neurological variations influence sensory perception. Demographic factors, including age, gender, and ethnicity, are significant in the study of synesthesia and neurological conditions. The distribution and type of synesthetic experiences vary across these demographic lines, suggesting that personal and cultural factors play a role in the manifestation and reporting of these unique sensory experiences.

Different synesthetic conditions lead to varied preferences in art and emotional responses. For instance, chromesthetes might prefer abstract art due to their experience of color in relation to sounds, while grapheme-color synesthetes may be drawn to certain visuals that align with their color associations with letters or numbers. This diversity in art appreciation underscores the profound impact of synesthetic perceptions on personal and emotional connections to art. The presence of multiple synesthetic types or combinations of neurological conditions within individuals suggests a layered and complex sensory profile landscape. Understanding these co-occurrences is crucial for a comprehensive view of how individuals experience and interpret their sensory environments, impacting everything from learning styles to daily interactions. The study reinforces the subjective and individualized nature of synesthetic experiences. Even within specific synesthetic conditions, individuals report a wide range of experiences and intensities. This variability reflects the deeply personal aspect of how people perceive and interact with the world, emphasizing the importance of individualized consideration in educational, therapeutic, and artistic contexts.

To illustrate the findings of the study on sensory profiles, neurological conditions, and synesthetic experiences, Table 4.1 outlines the key points. There is a noticeable pattern of specific synesthetic types being more prevalent within certain neurological conditions. For instance, individuals with ASC showed a higher occurrence of time–space synesthesia. Age and gender play a significant role in the prevalence and type of synesthesia reported. For example, grapheme-color synesthetes were predominantly female. Different synesthetic conditions influence preferences for various art types. Chromesthetics often appreciate digital and abstract art, highlighting their visual and auditory crossover experiences. Each group has distinct interactions with art and daily life due to their synesthetic experiences. For some, it enhances creativity and appreciation, while for others, it leads to sensory challenges and unique coping strategies. Thus, Table 4.1 represents a simplified view of complex interactions and should be interpreted with an understanding of the variability and personal nuances within each condition and synesthetic experience.

The findings from this section provide an insightful exploration into the diverse sensory worlds shaped by neurological conditions, synesthesia, and demographic factors. This exploration deepens our understanding of the myriad ways individuals perceive and engage with their environment,

Table 4.1 Sensory profiles, neurological conditions, and synesthetic experiences

Neurological condition	*Synesthesia type*	*Demographic consideration*	*Art appreciation*	*Key takeaways*
ADHD	Chromesthesia	Mostly adult, mixed gender	Digital, Abstract	High creativity, frequent art interaction, vivid experiences
Autism Spectrum Condition (ASC)	Time–Space	Adult, Male Dominance	Mixed Media, Digital	Intense emotional reactions, spatial visualization of time
Dyslexia	Lexical-Gustatory	Mostly Female, varied age	Abstract, Mixed Media	Enhanced memory, immersion in art, taste-texture sensations
PTSD	Mirror-Touch	Varied age, mostly Female	Digital, Tactile	Sensory overload, avoidance of stimuli, unique tactile experiences
Other (Self-Identified)	Grapheme-Color	Diverse, mostly Caucasian	Abstract, Realistic	Strong color associations, varied emotional intensity
Non-Synesthetes	None	Broad demographic range	Realistic, Abstract	Standard appreciation, less vivid sensory experiences

highlighting the richness and complexity of human perception and cognition. As we conclude our examination of sensory profiles, we recognize the potential for these insights to inform more personalized and empathetic approaches in various domains, including education, therapy, and artistic expression. Moving forward into the next section, we transition from quantitative analyses to qualitative insights, focusing on interviews that shed light on the self-exploration of aesthetic preferences. This next section aims to complement and enrich our understanding of how individuals uniquely experience and interact with art and aesthetics, offering a more holistic view of the synesthetic experience.

4.4 Diagnostic Approaches in Sensory Profile Analysis

In exploring the diagnostic approaches to determining sensory profiles and aesthetic preferences, research has focused on understanding the neural responses and cognitive basis for such preferences. Budziak and Lampert (2023) emphasize that intensive and diagnostic salience are key determinants of aesthetic preference, contingent on the categorical status of stimuli. Similarly, Machingura et al. (2020) highlight the Adult Sensory Profile as a diagnostic tool for evaluating and designing interventions for sensory processing preferences based on Dunn's Model of Sensory Processing. This approach underscores the importance of understanding individual sensory processing patterns to tailor interventions effectively.

Aesthetic inclinations and sensory encounters are shaped not only by neurological and psychological factors, but also by cultural and habitual elements. For instance, Flath et al., (2019) found that reading and writing habits influence aesthetic preferences for facial and bodily profiles, varying across different cultures. This cultural component emphasizes the importance of considering demographic factors when analyzing sensory profiles. From a practical standpoint, the Thai Sensory Profile Assessment Tool (TSPA) demonstrates the feasibility of classifying individuals for Mind–Body intervention based on their sensory preferences, indicating the potential for health promotion modalities tailored to individual sensory profiles (Kanchanawong et al., 2022). This finding points toward the significance of sensory diagnostic tools in personalizing healthcare and therapeutic practices.

Most recently, studies have used event-related potential (ERP) measurements to investigate the relationship between aesthetic preferences for design and neural responses (Cheng et al., 2023). Additionally, research has employed the production method to study aesthetic preference for sequences of human body postures, aiming to understand the cognitive basis for the development of aesthetic preference (Li et al., 2023). Furthermore, there is a growing interest in the use of empirical aesthetics to provide evidence-based and reproducible answers to questions related to aesthetic preferences (Trübutschek et al., 2023).

These studies reflect a multidisciplinary approach, integrating cognitive science, psychology, and neuroscience to gain a comprehensive understanding of aesthetic preferences. The multidimensional nature of sensory profiles and aesthetic preferences need to be considered, which are shaped by a confluence of diagnostic salience, cultural factors, and individual sensory processing patterns. These insights pave the way for more nuanced diagnostic approaches and interventions that consider the unique sensory world of individuals, ultimately contributing to more effective education, therapy, and health promotion strategies.

4.4.1 Exploratory and Interpretive Approach to Interviews

The study incorporated a qualitative follow-up through interviews to the initial quantitative survey, aiming to provide an in-depth exploration of the experiences of individuals with diverse neurological and synesthetic conditions. Those who completed the survey were invited for a 30-minute interview to discuss their sensory experiences, aesthetic preferences, and perceptions in greater detail. The objectives of the interviews were multifaceted. Firstly, they aimed to provide a platform for participants to articulate their sensory experiences and aesthetic preferences, enhancing understanding of their unique sensory profiles. Secondly, the interviews assessed the effectiveness of the questionnaire in helping participants understand their psychological, neuropsychological, and neurophysiological conditions, thus evaluating the survey's efficacy. Lastly, the goal was to utilize the insights gained to tailor ideal sensory experiences for individuals, potentially informing personalized approaches in therapeutic, educational, or recreational settings.

The methodology employed a careful selection of participants from the survey respondents, ensuring a diverse representation of neurological and synesthetic profiles. The interviews were semi-structured, balancing open-ended questions with more guided inquiries to thoroughly explore each participant's sensory world. The responses were meticulously transcribed and analyzed using thematic analysis to identify prevalent themes and patterns, enriching the quantitative data from the survey with qualitative depth. Confidentiality and consent were of utmost importance. All interviews were conducted with rigorous adherence to privacy standards, and participants were thoroughly informed of their rights, including the option to withdraw from the study at any time.

These interviews were pivotal in uncovering the subjective and personal nuances of sensory experiences and aesthetic preferences. By integrating quantitative and qualitative methodologies, the study aimed to offer a comprehensive understanding of how various neurological conditions and synesthetic experiences influence individual perceptions and interactions with the world. The insights derived from these interviews are anticipated to significantly advance personalized and sensitive approaches in a range of contexts, from designing therapeutic interventions to creating sensory-friendly environments.

The interview process was structured to provide a comprehensive exploration of the sensory profiles of participants, synesthetic experiences, and aesthetic preferences. The interviews began by establishing their diagnostic background and synesthesia type, if applicable. This foundational understanding guided the subsequent questions and discussion.

1. **Diagnostic Criteria and Synesthesia Type:** Participants were first asked about any diagnosed neurological conditions or self-diagnosed conditions, aiming to understand the context of their sensory experiences. They were then queried about the type of synesthesia they experience, if any. This helped in categorizing their experiences and understanding the specific sensory crossovers they encountered.
2. **Preferences and Dislikes:** The questionnaire delved into various sensory dimensions, including lighting, color, smell, sound, and texture, as well as the broader environment. Participants were asked to describe what they were drawn to or repelled by in each of these categories.

   Preferences: Questions were framed to uncover the types of art, environmental factors, sounds, and smells that participants found pleasurable or engaging.
   Dislikes: Conversely, participants were also asked about sensory experiences or environmental factors they found discomforting or distressing, providing a contrast to their preferences.

3. **Ideal Environment:** Based on their preferences and dislikes, participants were then asked to describe their ideal environment or setting that would elicit pleasure and comfort. This included a discussion on the optimal sensory conditions for their well-being, incorporating aspects of art, sound, texture, and more.
4. **Identifying Patterns in Artistic Preferences:** The interview aimed to identify any recurring patterns or themes in artistic preferences, considering their neurological or synesthetic conditions. This involved discussing their engagement with different forms of art and the sensory experiences associated with them.

5. **Diagnostics Through Artistic Engagement:** The final part of the interview explored how participants' interactions with art might serve as a diagnostic tool or therapeutic avenue. This included discussing how engagement with specific types of art or sensory environments might inform a better understanding of their condition or provide relief.

Throughout the interview, the questions (Table 4.2) were adapted and tailored based on the individual's responses, allowing for a fluid and organic conversation that respected the unique sensory world of each participant. The aim was to create a rich, qualitative data set that, when analyzed, would reveal insights into the sensory lives of individuals with different neurological and synesthetic profiles, informing more empathetic and personalized approaches in various fields. The organization of the interviews by condition, with a focus on synesthesia last, ensured a structured approach to analyzing and understanding the complex interplay of sensory experiences, preferences, and environmental factors.

The first interview was with a 33-year-old African-American female synesthetic interdisciplinary artist with chromesthesia, currently residing in New York City. She shared her lifelong journey with sound-color synesthesia, describing how it has shaped her artistic and professional endeavors. From an early age, she experienced colors when listening to music, such as washes of oranges and browns while listening to Michael Jackson's *Billie Jean* (1982). This multisensory perception was initially confusing but became a core part of her identity, especially after learning that others, like Pharrell Williams, shared similar experiences. It was upon learning of the experiences of other individuals she said: "Then realized that I'm not crazy and this happened to others."

Her artistic journey is deeply intertwined with her synesthetic experiences. As she notes: "I'm curious about how color and the senses amplify human experiences. As a synesthete with Chromesthesia (the ability to see sounds), I'm interested in how visuals and sound make us feel and is used as a mode of expression and connection through multisensory projects." As an engineer skilled in React, Vue, Rails, HTML, CSS, she has led teams in creating user experiences that illuminate the connection between product and user. She produces sound pieces and paints what she sees, creating artworks that represent the auditory visualizations she experiences. Her work aims to amplify the multisensory human experience through collaborative, integrity-driven practices, which has evolved

Table 4.2 Interview questionnaire

Aspect	*Details*
Diagnosis	
Synesthesia Type	
Preferences	
Art	
Environment	
Lighting	
Color	
Smell	
Sound	
Texture	
Environment (if applicable)	
Dislikes (Same as preferences but for dislikes)	
Art	
Environment	
Lighting	
Color	
Smell	
Sound	
Texture	
Environment (if applicable)	
Ideal Environment (Based on answers)	

into a codex of symbols representing different musical elements, such as burgundy circles for drum kicks or cobalt blue lines for bass D notes. This personal language allows her to translate the music she hears into visual art. Her works range from interpreting the music of others to create her own, noticing a distinct difference in the visual output when the music is internally versus externally sourced.

She mentioned exploring immersive exhibitions but prefers in-person and analog experiences that allow for a full sensory immersion, indicative of her need for an elevated sensory experience due to overlapping neurologies. She described entering a "flow state" when deeply engaged with music, a state of intense focus and sensory integration that is involuntary and deeply immersive. A flow state, also known as "being in the zone," is a mental state in which a person performing an activity is fully immersed in a feeling of energized focus, full involvement, and enjoyment (Jackson & Marsh, 1996).

Her preferred art style is abstract, noting specific experiences with chromaesthetic reactions to artwork, such as Leon Berkowitz's (1911–1987) Coronation (1979) at the National Gallery of Art in the Corcoran Collection in Washington, D.C., where she heard a buzzing sound emanating from the pastel sunset colors of the painting. However, she expressed a dislike for certain immersive art installations like that at the Natural History Museum in DC that create dissonance between her synesthetic perceptions and the intended sensory outputs.

In discussing her professional life, she emphasized the importance of multisensory experiences in branding and designing, citing her work and inspirations from others in the field who leverage sensory experiences to create impactful and healing environments. As a UX engineer and creative technologist, she continues to explore the intersection of art, technology, and sensory experiences, crafting spaces and products that resonate on a multisensory level with audiences.

Alongside the first participant with chromesthesia, a 57-year-old Caucasian male synesthete with the same condition shared his unique synesthetic journey. He discovered his synesthesia at 56, around the same time he embraced his gay identity, leading him to question the relationship between his newfound identity and enhanced sensory experiences. He describes vivid, uncontrollable physical reactions to sound that commenced soon after his self-acknowledgment of being gay, likening these episodes to seizures and intense physical sensations like sexual intercourse and kept asking himself "Why am I suddenly feeling sound?" Along with the ability to see sound was a newfound set of creative abilities, including singing and dancing that have since been externally validated by professionals in the field. He believes this discovery led to a broader neural network engagement and a deeper understanding of his identity and sensory experiences. As he continually noted about his journey of self-discovery and of broader investigation of the human condition: "When I look at people, I ask who validated you?".

This participant delves into the neurological aspects of his condition, discussing the role of the ventral tegmental area (VTA) of the brain in dopamine production and its relationship with sensory inputs and emotions (Trutti et al., 2019). He redefines ADHD as "Neuroconvergence," suggesting that everyone has a degree of neurodiversity, and it is the interpretation and framing of sensory inputs and feelings that shape our emotional world. He describes his brain as entering dopamine loops during periods of mania, enhancing his intelligence and emotional depth perception. For him, synesthesia is not just an isolated phenomenon but a repetitive interaction with stimuli that deeply influences his daily life and perception of the world.

With a background in Decision Science, he explains how his early engagement with complex systems and programming shaped his understanding of himself and the world around him. He believes in stepping outside oneself to gain a comprehensive understanding, a method he applies to his synesthetic experiences and his life philosophy. His narrative provides a poignant example of how neurological conditions, personal identity, and sensory experiences intersect to shape one's interaction with the world, emphasizing the complexity and personal nature of synesthetic experiences.

Following these synesthetic conditions, the study further explored physiology of sight and color in the case of a 57-year-old Caucasian female with confirmed Tetrachromacy. This rare condition enables her to perceive a broader spectrum of colors than the typical trichromatic vision found in most humans. Her sensory experience is not only expansive but also deeply integrated with her personal and creative life. She recounts her first awareness of her unique perception in kindergarten, where her precise and vivid descriptions of color extended beyond the usual vocabulary, drawing from the rich nomenclature of Crayola crayons (e.g. instead of saying blue, she would say Cornflower Blue). Living in New York City, she has leveraged her extraordinary visual ability into a career as an avid bird watcher and photographer, capturing the subtle nuances of color that many might miss.

The diagnosis of Tetrachromacy came later in life through DNA testing, revealing the potential for an enhanced color spectrum, albeit not necessarily functional. Despite not having other neurodivergent diagnoses, she often experiences migraines, attributing them to the overwhelming sensory stimulation her condition brings. Her synesthetic experience extends to vivid disagreements over color perceptions, with a particularly profound memory aid benefit. She speculates on the potential evolutionary advantages of synesthesia, linking it to heightened sensory awareness. This connection sparks a thought in her about the possible practical applications of her heightened color perception, such as detecting health changes or conditions through visual cues.

In her daily life, the participant expressed a preference for photorealistic art and macrophotography, resonating with her ability to see and appreciate minute color details. She imagines an ideal art exhibit that would allow others to understand the depth of her color perception, proposing an anteroom with paint chips representing the full spectrum she experiences. However, abstract art tends to make her uncomfortable, despite acknowledging its beauty. Her ideal environment is one bathed in natural or candlelight, avoiding the harshness of artificial lighting. She finds solace in the softness of down feathers, the melody of birdsongs, and the fragrance of florals. As someone with hyper-sensory experiences, she emphasizes the need for gentleness in her surroundings to mitigate the intensity of her perceptions.

We now transition to look at how neurological conditions affect sense experience with two cases. The personal experiences of a 38-year-old Caucasian female and a 43-year-old Caucasian male, both with dyslexia, provide a fascinating glimpse into how neurological differences shape sensory perception and aesthetic preferences. The 38-year-old woman's diagnosis of dyslexia came later in life, a common narrative for many adults who went through schooling before dyslexia was widely understood or effectively screened. Her journey resonates with the struggle of re-reading, comprehension difficulties, and the anxiety that comes with such challenges. Her creative outlet through felted sculptures and the use of Midjourney to create a book centered around self-regulation for children is a testament to her adaptability and desire to contribute positively, leveraging her unique sensory experiences. Her aesthetic preferences lean toward tactile art forms like felt, where the physical act of needle felting is both a creative and cathartic process. She is drawn to color, needing the vibrancy and variety it offers, except for certain shades she misidentifies due to color vision peculiarities. Her ideal space is minimalist and open, with white walls acting as a blank canvas against which any piece becomes the focal point.

On the other hand, the 43-year-old man identifies as dyslexic with a keen ability to skim large datasets and understand overarching themes and patterns. This skill, though beneficial in certain contexts, makes the focused task of reading and comprehending individual lines of text quite challenging. His synesthetic experience seems to complicate his interaction with art, as he views pieces holistically, often missing out on the details that others might point out. His connection to nature is evident in his preference for sage green and nature scenes, which provide a sense of peace and familiarity. He seeks out environments that are dimly lit or dark, as brightness can trigger migraines—a condition he has struggled with since childhood. His office, a carefully curated space with just enough light to differentiate colors without overwhelming his senses, reflects his need for a controlled, comfortable environment.

Both individuals illustrate how dyslexia, coupled with unique sensory processing and aesthetic preferences, shapes their interaction with the world. Their stories underscore the importance of understanding and accommodating neurodiversity in all its forms, as each person navigates their environment in a way that brings comfort and meaning to their lives. The insights gained from these interviews contribute to a broader understanding of how sensory profiles are influenced by neurological conditions and how these profiles, in turn, influence one's interaction with art and their surrounding environment.

Before turning to our last case study involving technological interventions, we look at the case of a 56-year-old Caucasian female with time–space synesthesia, which offers a unique perspective on sensory experiences and preferences. Identifying as gay and potentially neurodivergent, she has navigated life with sensitivities to loud, busy environments, and certain textures that resonate with her daughter's autism diagnosis. This personal connection has led to an unmasking process, as she recognizes and relates more to her child's experiences and the broader neurodivergent community. Moreover, her professional role as a department chair has benefited from her synesthetic ability to visualize time in three dimensions, an organizational skill she thought was common until realizing its uniqueness. She describes her experience of time as akin to an organized, moving shoe closet, enabling her to manage schedules and tasks effectively.

Her aesthetic preferences are for immersive, large-scale art and architecture that fills a room and allows her to be enveloped in the experience. She feels a sense of awe in places like the Hagia Sophia or the Vatican, preferring environments where the art dominates and creates an atmosphere of grandeur and contemplation. While she has no particular preference for open or closed spaces in general, she enjoys cozy settings and is comfortable being alone in vast natural landscapes or in small gatherings with close friends. Despite her appreciation for immersive art, she avoids spaces with too many pieces or busy street fairs, finding them either overwhelming or lacking the focus she desires. Her ideal environment includes sunlight and windows, muted colors, and minimal clutter, providing a serene and undecorated space for relaxation and thought. Reading lamps, bookshelves, and her dog are essential elements of her bespoke space, creating an atmosphere conducive to her well-being.

Interestingly, art museums stand out as the most "energizing and uplifting" places for her, offering solitude and inspiration. She values the ability to explore and contemplate art independently, free from external pressures or influences. Libraries also hold a special place as a quiet, contemplative environment where she can engage in "parallel play," enjoying the presence of others while immersed in her own thoughts. This case study highlights how sensory profiles, shaped by synesthesia and potential neurodivergence, influence not only aesthetic preferences and comfort in various environments but also personal and professional life.

The last case studies are those of neurophysiological or neuroacquired conditions. These individuals suffered different forms of physical trauma that led to or compounded their sense perception. The first is that of a 34-year-old Caucasian male grappling with photophobia and Boday syndrome after a battle with testicular cancer. The narrative provides a deep insight into the lived experience of sudden visual impairment and its cascading effects on life. Initially an outdoorsy person deeply engaged with technology, his life took a dramatic turn when he lost central vision post-chemotherapy at 21. The gradual and then rapid loss of sight and the subsequent development of photophobia significantly altered his daily life, leading to years of navigating medical interventions and personal adjustments.

The journey of this individual is marked by a continuous struggle with the medical system and his efforts to understand and adapt to his new sensory world. The surgical interventions, including an eight-hour surgery leading to further vision loss and the eventual removal of his eye, highlight the severity of his condition and the complexities involved in treating such cases. His experiences with light and dark vision, the pain of photophobia, and the eventual adaptation to life with significant visual impairment underscore the resilience and adaptability of the human spirit.

The narrative shared by the subject also delves into the importance of support systems and technology for individuals with disabilities. The use of various technological aids, such as the IrisVision VR headset, OrCam, and customized home automation systems, reflects a proactive approach to adapting his environment to his needs. As he noted of himself and the students he works with, it is important to "show them that the world can bend to them instead of them breaking for the world." These technologies not only facilitate daily tasks but also empower him to continue his work and hobbies with a degree of independence. His insights into the perception and societal understanding of visual impairment shed light on the challenges faced by those living with disabilities. From the need for more nuanced and accessible technologies to the importance of societal support and understanding, his story is a call to action for a more inclusive and accommodating world.

The reflections on his condition, support systems, and the use of technology to adapt and thrive provide a powerful testament to the human capacity to overcome adversity. His story adds a valuable perspective to the understanding of visual impairment, the importance of personalized accommodations, and the potential of technology to transform lives. As noted, "Anyone who is different strives to be normal and anyone who is normal strives to be different." His journey from fully sighted to navigating the world with significant visual impairment is a poignant example of the complex interplay between the human sensory experience, technology, and the resilience of the human spirit.

The second, and last, case study is of a 61-year-old Caucasian female who has a range of experiences and conditions that developed over the course of her life that shape her sensory world. Diagnosed with ADHD, she has always needed something to focus on to remain still. Born blind, she underwent surgery at a year and a half old and received her first pair of glasses, introducing her to a world of color that would become central to her life. Her childhood was marked by a unique perception of the world, nurtured by her initial blindness. Her imagination ran wild, feeling her mother's face and tracing it, marveling at the colors in a single blade of grass. This early connection to sensory experiences led her to art in the 6th grade, under the tutelage of Helen Price Stacy. Her journey through art became a healing process, helping her navigate and express the complexities of her experiences, including her service in the Navy and the subsequent traumas she endured.

These neuroacquired conditions stem from an injury in the Navy, leaving her with a broken back and with additional trauma leading to PTSD from sexual assault. Since then, art serves as a replenishment for her, a never-ending process that helps her cope with stress and intensifies her experience of color. The intensity of her color experience is such that lying on grass and gazing at the clouds brings her calm, with blue being particularly soothing. Her synesthesia manifests as hypersensitivity to sound, with certain noises causing painful reactions that resonate through her body. She prefers bright, neon colors and finds equilibrium in organic balance, noting that any imbalance is immediately felt. Her artistic preferences lean toward uniformity in art and music, favoring both abstract and representational forms. She dislikes imbalance, white walls, dark colors, and the texture of velvet. Her ideal environment is one of stillness and quiet, with off-white walls, allowing her to gaze at clouds and let her imagination roam, aiding her breathing and providing a much-needed brain break. The story of this subject is a testament to the power of art as a sanctuary and a tool for healing and underscores the importance of understanding individual sensory profiles and preferences, especially for those with complex neurodiverse conditions. Her experiences illuminate the potential for tailored sensory experiences to enhance well-being and offer solace in a world that can often be overwhelming.

As these case studies have demonstrated, the exploration of sensory profiles through the lens of synesthesia and various neurological conditions reveals a diversity of individual experiences and preferences. This chapter has illuminated the diverse ways individuals interact with and perceive their environment, influenced by a range of factors from genetic conditions like tetrachromacy to acquired conditions like low vision. The narratives of those with chromesthesia, time–space synesthesia, dyslexia, and other conditions provide a glimpse into the unique sensory worlds individuals navigate daily, shaping their preferences, dislikes, and ideal environments.

Key takeaways from this chapter include the recognition of the broad spectrum of sensory experiences and the importance of understanding these differences for creating inclusive and supportive environments. Synesthesia, while often considered in the realm of the extraordinary, provides valuable insights into the human experience of perception and cognition. It underscores the varied ways the brain can interpret sensory information, leading to unique experiences such as seeing sounds or visualizing time as spatial structures. Moreover, the individual stories highlight the need for empathy and personalized approaches in education, therapy, and design. Recognizing that each person's sensory experience can significantly differ, there is a need for flexible and adaptable environments that cater to a wide range of sensory needs and preferences.

Technology emerges as a powerful tool in facilitating these adaptive environments, as evidenced by the innovative use of visual aids, specialized software, and sensory devices in supporting individuals with different sensory processing needs. Advancements in assistive technology improve independence and quality of life for individuals with sensory impairments, while also providing innovative solutions for enriching the interaction of all individuals with their surroundings in more profound ways. As we progress, the concluding chapter will explore the future of neuroarts and its potential to craft personalized sensory experiences. This emerging field holds the promise to further bridge the divide between sensory encounters and individual requirements, introducing tailored augmenting comprehension, accessibility, and enjoyment of the surrounding world. By drawing upon insights from synesthesia, neurological research, and technological advancements, the future of neuroarts has the capacity to redefine how we design environments, create art, and engage with our surroundings in ways that resonant with each person's distinct sensory profile.

References

Adams, K., Boué, S., & Watfern, C. (2022). Being together in a neurodiverse world exploring with project art works. *The Big Anxiety: Taking Care of Mental Health in Times of Crisis*, 239.

Adams, K. A., & McGuire, E. K. (2022). *Research methods, statistics, and applications.* Sage.

Ahvan, Y. R., Sheikhei Fini, A. A., & Zainalipor, H. (2022). The impact of the hidden curriculum components on cognitive characteristics of high school students' learning. *Journal of Fundamentals of Mental Health.* https://doi.org/10.22038/JFMH.2022.19442

Andreev, S. S., Moskvoretskiy, V. V., Gorin, A. A., & Zinchenko, O. O. (2023). Grapheme-color synesthesia induction with V4 transcranial direct current stimulation (No. WP BRP 137/PSY/2023). *National Research University Higher School of Economics.*

Aoki, Y., Shibasaki, M., & Nakata, H. (2023). Synesthesia has specific cognitive processing during Go/No-go paradigms. *Scientific Reports, 13*(1), 6172.

Ayton, D. (2023). Phenomenology. In *Qualitative Research—A practical guide for health and social care researchers and practitioners.* Monash University.

Bankieris, K., & Simner, J. (2014). Sound symbolism in synesthesia: Evidence from a lexical–gustatory synesthete. *Neurocase, 20*(6), 640–651.

Banks, A. L. (2022). *Self-perceived social communication perspectives of autistic young adults: a phenomenological qualitative study.*

Banissy, M. J., Garrido, L., Kusnir, F., Duchaine, B., Walsh, V., & Ward, J. (2011). Superior facial expression, but not identity recognition, in mirror-touch synesthesia. *Journal of Neuroscience, 31*(5), 1820–1824.

Bernett, S. (2022). *Composing with the event—Moving toward neurodiverse perception/sensation* (Doctoral dissertation, Concordia University).

Bolwerk, A., Mack-Andrick, J., Lang, F. R., Dörfler, A., & Maihöfner, C. (2014). How art changes your brain: Differential effects of visual art production and cognitive art evaluation on functional brain connectivity. *PLoS ONE, 9*(7), e101035.

Borisova, Y., Sharapova, D., Turayev, B., & Shernazarov, F. (2023). Concomitant mental disorders and social functioning of adults with high-functioning autism/asperger syndrome. *Science and Innovation, 2*(D11), 36–41.

Bouzy, J., Brunelle, J., Cohen, D., & Condat, A. (2023). Transidentities and autism spectrum disorder: A systematic review. *Psychiatry Research, 323*, 115176.

Bowery, A. M. (2007). Know thyself: Socrates as storyteller. In *Philosophy in dialogue: Plato's many devices* (pp. 82–110).

Brang, D., & Ahn, E. (2019). Double-blind study of visual imagery in grapheme-color synesthesia. *Cortex, 117*, 89–95.

Brown, S., Gao, X., Tisdelle, L., Eickhoff, S. B., & Liotti, M. (2011). Naturalizing aesthetics: Brain areas for aesthetic appraisal across sensory modalities. *NeuroImage, 58*(1), 250–258.

Budziak, J., & Lempert, D. (2023). Aesthetic preferences and policy preferences as determinants of US Supreme Court writing style. *Journal of Law and Courts, 11*(1), 45–66.

Camp, A., Pastrano, A., Gomez, V., Stephenson, K., Delatte, W., Perez, B., Syas, H., & Guiseppi-Elie, A. (2021). Understanding ADHD: Toward an innovative therapeutic intervention. *Bioengineering, 8*(5), 56.

Carmichael, D. A., Down, M. P., Shillcock, R. C., Eagleman, D. M., & Simner, J. (2015). Validating a standardised test battery for synesthesia: Does the Synesthesia Battery reliably detect synesthesia? *Consciousness and Cognition, 33*, 375–385.

Cheng, Y., Chen, J., Li, J., Li, L., Hou, G., & Xiao, X. (2023). Research on the preference of public art design in urban landscapes: Evidence from an event-related potential study. *Land, 12*(10), 1883.

Chiou, R., & Rich, A. N. (2014). The role of conceptual knowledge in understanding synaesthesia: Evaluating contemporary findings from a "hub-and-spokes" perspective. *Frontiers in Psychology, 5*, 105.

Christias, D. (2023). Lifeworld phenomenology after Husserl: Merleau-Ponty, enactivism, Heidegger and science. In *Normativity, lifeworld, and science in Sellars' synoptic vision* (pp. 63–89). Springer Nature Switzerland.

Churchill, W., & Wilde, O. (2022). The creative trance and the brain. In *The creative trance: Altered states of consciousness and the creative process* (p. 80).

Cid Rodriguez, M. J., Ribes, R., & Janku, K. (2022). Sensory profile in the Snoezelen intervention. *Social Pathology and Prevention, 7*(1), 35–47.

Cohen, S. R., Joseph, K., Levinson, S., Blacher, J., & Eisenhower, A. (2022). "My autism is my own": Autistic identity and intersectionality in the school context. *Autism in Adulthood, 4*(4), 315–327.

Cunliffe, L., Coulthard, H., & Williamson, I. R. (2022). The lived experience of parenting a child with sensory sensitivity and picky eating. *Maternal & Child Nutrition, 18*(3), e13330.

Cytowic, R. E. (2002). *Synesthesia: A union of the senses.* MIT press.

de Giambattista, C., Ventura, P., Trerotoli, P., Margari, M., Palumbi, R., & Margari, L. (2019). Subtyping the autism spectrum disorder: Comparison of children with high functioning autism and Asperger syndrome. *Journal of Autism and Developmental Disorders, 49*(1), 138–150.

Di Paolo, E. A. (2023). F/acts ways of enactive worldmaking. *Journal of Consciousness Studies, 30*(11–12), 159–189.

Eagleman, D. M., Kagan, A. D., Nelson, S. S., Sagaram, D., & Sarma, A. K. (2007). A standardized test battery for the study of synesthesia. *Journal of Neuroscience Methods, 159*(1), 139–145.

Evans, J. A., Krumrei-Mancuso, E. J., & Rouse, S. V. (2023). What you are hiding could be hurting you: Autistic masking in relation to mental health, interpersonal trauma, authenticity, and self-esteem. *Autism in Adulthood*. https://doi.org/10.1089/aut.2022.0115

Fellowes, S. (2023). Self-diagnosis in psychiatry and the distribution of social resources. *Royal Institute of Philosophy Supplements, 94*, 55–76.

Flath, M. E., Smith, A. K., & Elias, L. J. (2019). Cultural differences in lateral biases on aesthetic judgments: The effect of native reading direction. *Culture and Brain, 7*, 57–66.

Gallagher, S. (2022). What is phenomenology? *Phenomenology* (pp. 1–10). Springer International Publishing.

Golan, O., Haruvi-Lamdan, N., Laor, N., & Horesh, D. (2022). The comorbidity between autism spectrum disorder and post-traumatic stress disorder is mediated by brooding rumination. *Autism, 26*(2), 538–544.

Hagtvedt, H. (2022). A brand (new) experience: Art, aesthetics, and sensory effects. *Journal of the Academy of Marketing Science, 50*(3), 425–428.

Hase, A., Haynes, M., & Hasler, G. (2023). Using simple economic games to assess social orienting and prosocial behavior in adolescents with autism spectrum disorder. *Autism Research, 16*, 1199–1209.

Hauw, F., El Soudany, M., Rosso, C., Daunizeau, J., & Cohen, L. (2023). A single case neuroimaging study of tickertape synesthesia. *Scientific Reports, 13*(1), 12185.

Hiscox, L. V., Bray, S., Fraser, A., Meiser-Stedman, R., Seedat, S., & Halligan, S. L. (2023). Sex differences in the severity and natural recovery of child PTSD symptoms: A longitudinal analysis of children exposed to acute trauma. *Psychological Medicine, 53*(6), 2682–2688.

Hitch, D., Wilson, C., & Hillman, A. (2023). Sensory modulation in mental health practice. *Mental Health Practice, 26*(6).

Hoffman, S. N., Urosevich, T. G., Kirchner, H. L., Boscarino, J. J., Dugan, R. J., Withey, C. A., ... & Boscarino, J. A. (2019). Grapheme-color synesthesia is associated with PTSD among deployed veterans: confirmation of previous findings and need for additional research. *International Journal of Emergency Mental Health, 21*(1).

Hu, X. Z., Ursano, R. J., Benedek, D., Li, X., Li, Z., Zhang, L., & Biomarker Study Group. (2023). Sex-related difference of association of mitochondrial DNA copy number with PTSD in US service members. *Journal of Psychiatric Research, 159*, 1–5.

Hubbard, E. M. (2007). Neurophysiology of synesthesia. *Current Psychiatry Reports, 9*(3), 193–199.

Hupé, J. M., & Dojat, M. (2015). A critical review of the neuroimaging literature on synesthesia. *Frontiers in Human Neuroscience, 9*, 103.

Hutson, P., & Hutson, J. (2023a). Neurodiversity and inclusivity in the workplace: Biopsychosocial INTERVENTIONS FOR PROMOTING COMPETITIVE ADVAntage. *Journal of Organizational Psychology, 23*(2), 1–16.

Hutson, J., & Hutson, P. (2023b). *Sensemaking and neuroaesthetics study.* Lindenwood University.

Hutson, P., & Hutson, J. (2023c). Autism in females: Understanding the overlooked diagnoses, unique challenges, and recommendations. *Journal of Clinical & Medical Images, 3*(3). https://doi.org/10.55920/2771-019X/1461

Jackson, S. A., & Marsh, H. W. (1996). Development and validation of a scale to measure optimal experience: The Flow State Scale. *Journal of Sport and Exercise Psychology, 18*(1), 17–35.

Jacobi, C. J., Varga, P. J., & Vaidyanathan, B. (2022). Aesthetic experiences and flourishing in science: A four-country study. *Frontiers in Psychology, 13*, 923940.

Johnson, V. E., Courtney, K., & Chng, K. (2023). Race-based traumatic stress predicts risky drinking, over and above negative affect, and non-race-related trauma symptoms in racial/ethnic minority female college students. *Journal of Racial and Ethnic Health Disparities*, 1–11.

Kanchanawong, T., Prasoetsang, T., Limvongvatana, F., Ooraikul, L., Kunapun, S., Srikhamjak, T., & Gomutbutra, P. (2022). The feasibility of the Thai Sensory Profile Assessment Tool (TSPA) for classifying the participants for mind-body intervention. *medRxiv, 2022-06.*

Kinnear, D., Rydzewska, E., Dunn, K., Hughes-McCormack, L., Melville, C., Henderson, A., & Cooper, S. A. (2020). The relative influence of intellectual disabilities and autism on sensory impairments and physical disability: A whole-country cohort of 5.3 million children and adults. *Journal of Applied Research in Intellectual Disabilities, 33*(5), 1059–1068.

Kliem, E., Forster, M., & Leder, H. (2022). Aesthetic preference for negatively-valenced artworks remains stable in pathological aging: A comparison between cognitively impaired patients with Alzheimer's disease and healthy controls. *Frontiers in Psychology, 13*, 879833.

Kyzar, E. J., & Denfield, G. H. (2023). Taking subjectivity seriously: Towards a unification of phenomenology, psychiatry, and neuroscience. *Molecular Psychiatry, 28*(1), 10–16.

Lebeau, C., & Richer, F. (2022). Emotions and consciousness alterations in music-color synesthesia. *Auditory Perception & Cognition, 5*(1–2), 76–85.

Lebrasseur, A., Fortin-Bédard, N., Lettre, J., Bussières, E. L., Best, K., Boucher, N., ... & Routhier, F. (2021). Impact of COVID-19 on people with physical disabilities: a rapid review. *Disability and Health Journal, 14*(1), 101014.

Li, J., Wu, S. R., Zhang, X., Luo, T. J., Li, R., Zhao, Y., Liu, B., & Peng, H. (2023). Cross-subject aesthetic preference recognition of Chinese dance posture using EEG. *Cognitive Neurodynamics, 17*(2), 311–329.

Licciardi, L., & Brown, T. (2023). An overview & critical review of the sensory profile–second edition. *Scandinavian Journal of Occupational Therapy, 30*(6), 758–770.

Lin, Q., Shi, Y., Huang, H., Jiao, B., Kuang, C., Chen, J., ... & Ma, L. (2023). Functional brain network alterations in the co-occurrence of autism spectrum disorder and attention deficit hyperactivity disorder. *European Child & Adolescent Psychiatry*, 1–12.

Machingura, T., Kaur, G., Lloyd, C., Mickan, S., Shum, D., Rathbone, E., & Green, H. (2020). An exploration of sensory processing patterns and their association with demographic factors in healthy adults. *Irish Journal of Occupational Therapy, 48*(1), 3–16.

Malik, S. (2022). Using neuroscience to explore creative media in art therapy: A systematic narrative review. *International Journal of Art Therapy, 27*(2), 48–60.

Mattheisen, M., Grove, J., Als, T. D., Martin, J., Voloudakis, G., Meier, S., ... & Børglum, A. D. (2022). Identification of shared and differentiating genetic architecture for autism spectrum disorder, attention-deficit hyperactivity disorder and case subgroups. *Nature Genetics, 54*(10), 1470–1478.

Meier, B., & Rothen, N. (2013). Grapheme-color synaesthesia is associated with a distinct cognitive style. *Frontiers in Psychology, 4*, 632.

Merten, N., Pinto, A. A., Paulsen, A. J., Chen, Y., Dillard, L. K., Fischer, M. E., ... & Cruickshanks, K. J. (2022). The association of psychological well-being with sensory and cognitive function and neuronal health in aging adults. *Journal of Aging and Health, 34*(4–5), 529–538.

Meyer, R., & Brancazio, N. (2022). Putting down the revolt: Enactivism as a philosophy of nature. *Frontiers in Psychology, 13*, 948733.

Micai, M., Fatta, L. M., Gila, L., Caruso, A., Salvitti, T., Fulceri, F., ... & Scattoni, M. L. (2023). Prevalence of co-occurring conditions in children and adults with autism spectrum disorder: a systematic review and meta-analysis. *Neuroscience & Biobehavioral Reviews, 155*, 105436.

Miller, C. A., & Hübner, R. (2023). The relations of empathy and gender to aesthetic response and aesthetic inference of visual artworks. *Empirical Studies of the Arts, 41*(1), 188–215.

Moraru, C. (2023). The cultural category of neurodiversity: Art spaces for creating identities and understanding neurominorities. *Espace, 133*, 6–16.

Mowlem, F. D., Rosenqvist, M. A., Martin, J., Lichtenstein, P., Asherson, P., & Larsson, H. (2019). Sex differences in predicting ADHD clinical diagnosis and pharmacological treatment. *European Child & Adolescent Psychiatry, 28*, 481–489.

Neckar, M., & Bob, P. (2014). Neuroscience of synesthesia and cross-modal associations. *Reviews in the Neurosciences, 25*(6), 833–840.

Özer, H. (2019). A case who has smell-color synesthesia. *Selcuk Medical Journal, 35*(1), 55–57.

Panagiotidi, M., Overton, P. G., & Stafford, T. (2020). The relationship between sensory processing sensitivity and attention deficit hyperactivity disorder traits: A spectrum approach. *Psychiatry Research, 293*, 113477.

Pantazakos, T., & Vanaken, G. J. (2023). Addressing the autism mental health crisis: The potential of phenomenology in neurodiversity-affirming clinical practices. *Frontiers in Psychology, 14*. https://doi.org/10.3389/fpsyg.2023.1225152

Pawlett-Jackson, S. (2022). Many faces, plural looks: Enactive intersubjectivity contra Sartre and Levinas. *Phenomenology and the Cognitive Sciences, 21*(4), 903–925.

Pearson, J. (2019). The human imagination: The cognitive neuroscience of visual mental imagery. *Nature Reviews Neuroscience, 20*(10), 624–634.

Pesce, D., Neirotti, P., & Paolucci, E. (2019). When culture meets digital platforms: Value creation and stakeholders' alignment in big data use. *Current Issues in Tourism, 22*(15), 1883–1903.

Pezzulo, G., Parr, T., & Friston, K. (2022). The evolution of brain architectures for predictive coding and active inference. *Philosophical Transactions of the Royal Society B, 377*(1844), 20200531.

Redshaw, R., & McCormack, L. (2022). "Being ADHD": A qualitative study. *Advances in Neurodevelopmental Disorders, 6*(1), 20–28.

Rinaldi, L. J. (2021). Synesthesia and sensory processing. In *Oxford research encyclopedia of neuroscience.* Oxford University Press.

Sadeghzadeh, N., Farajzadeh, N., Dattatri, N., & Acevedo, B. P. (2023). SPS vision net: Measuring sensory processing sensitivity via an artificial neural network. *Cognitive Computation,* 1–14. https://doi.org/10.1007/s12559-023-10216-6

Sepúlveda-Pedro, M. A. (2023). *Enactive cognition in place: Sense-making as the development of ecological norms.* Springer Nature.

Sheehan-Klein, M. (2024). *A polyvagal theory and mindfulness based art therapy approach to self-regulation and self compassion in the face of trauma.*

Silberman, S. (2017). Neurotribes: The legacy of autism and the future of neurodiversity. *Psychiatry, 30*(2), 120–123.

Simner, J., Mulvenna, C., Sagiv, N., Tsakanikos, E., Witherby, S. A., Fraser, C., Scott, K., & Ward, J. (2006). Synaesthesia: The prevalence of atypical cross-modal experiences. *Perception, 35*(8), 1024–1033.

Smith, O., & Jones, S. C. (2020). 'Coming out'with autism: Identity in people with an Asperger's diagnosis after DSM-5. *Journal of Autism and Developmental Disorders, 50*, 592–602.

Stanek, C., & Mattson, G. (2023). 'At least you're not neurotypical': Stigma, mental illness disclosure, and social capital among privileged college students. *Deviant Behavior*, 1–20.

Steinbrenner, J. R., McIntyre, N., Rentschler, L. F., Pearson, J. N., Luelmo, P., Jaramillo, M. E., Boyd, B. A., Wong, C., Nowell, S. W., Odom, S. L., & Hume, K. A. (2022). Patterns in reporting and participant inclusion related to race and ethnicity in autism intervention literature: Data from a large-scale systematic review of evidence-based practices. *Autism, 26*(8), 2026–2040.

Thielen, H., Tuts, N., Lafosse, C., & Gillebert, C. R. (2023). The neuroanatomy of poststroke subjective sensory hypersensitivity. *Cognitive and Behavioral Neurology, 36,* 10–1097.

Tihanyi, B. T., Ferentzi, E., Beissner, F., & Köteles, F. (2018). The neuropsychophysiology of tingling. *Consciousness and Cognition, 58*, 97–110.

Trübutschek, D., Yang, Y. F., Gianelli, C., Cesnaite, E., Fischer, N. L., Vinding, M. C., ... & Nilsonne, G. (2023). EEGManyPipelines: A large-scale, grass-roots multi-analyst study of electroencephalography analysis practices in the wild. *Journal of Cognitive Neuroscience,* 1–8.

Trutti, A. C., Mulder, M. J., Hommel, B., & Forstmann, B. U. (2019). Functional neuroanatomical review of the ventral tegmental area. *NeuroImage, 191*, 258–268.

Van Leeuwen, T. M., Den Ouden, H. E., & Hagoort, P. (2011). Effective connectivity determines the nature of subjective experience in grapheme-color synesthesia. *Journal of Neuroscience, 31*(27), 9879–9884.

Vessel, E. A., Starr, G. G., & Rubin, N. (2012). The brain on art: Intense aesthetic experience activates the default mode network. *Frontiers in Human Neuroscience, 6*, 66.

Ward, J. (2013). Synesthesia. *Annual Review of Psychology, 64*, 49–75.

Weiss, P. H., Zilles, K., & Fink, G. R. (2005). When visual perception causes feeling: Enhanced cross-modal processing in grapheme-color synesthesia. *NeuroImage, 28*(4), 859–868.

Wendin, K., Pálsdóttir, A. M., Spendrup, S., & Mårtensson, L. (2023). Odor perception and descriptions of rose-scented geranium pelargonium graveolens 'Dr. Westerlund'–Sensory and Chemical Analyses. *Molecules, 28*(11), 4511.

Winner, E. (2019). *How art works: A psychological exploration*. Oxford University Press.

Woodbury-Smith, M. R., & Volkmar, F. R. (2009). Asperger syndrome. *European Child & Adolescent Psychiatry, 18*, 2–11.

PART IV

The Future of Neuroarts

CHAPTER 5

Conclusion: Future Applications, Technologies, and Directions

The following conclusion serves as a forward-looking summary of the evolving field of neuroarts. It discusses the potential for personalization in neuroaesthetic experiences through sensory augmentation and advanced neurological technologies. The chapter explores future applications, including virtual art in therapy, AI's role in sensory enhancement, and the use of tactile therapies for multisensory artistic engagement. It also delves into neurophysiological stimulation in art, such as affective computing and cognitive techniques, and examines the intersection of AI with artistic exploration, particularly in creating immersive, virtual experiences. Ethical considerations in neuroarts, including data privacy and the impact of commercialization, are highlighted, emphasizing the need for responsible development. The chapter emphasizes the significance of involving the public and incorporating neuroarts into educational and healthcare contexts, highlighting its potential as a therapeutic resource. It culminates by exploring the technological frontiers within the realm of neuroaesthetics, such as brain – computer interfaces and the utilization of machine learning in art generation. Additionally, it addresses theoretical inquiries regarding subjectivity and realism in art. In essence, this chapter portrays a field brimming with opportunities, poised to reshape our comprehension and interaction with art by leveraging the insights of neuroscience and technology.

J. Hutson et al., *Sensemaking and Neuroaesthetics*,
https://doi.org/10.1007/978-3-031-58045-1_5

5.1 Potential for Further Personalization in Neuroarts

Concluding the insights garnered from the diverse sensory profiles and experiences of synesthetes in Chapter 4, it becomes clear that the evolving landscape of emerging technologies, including extended reality (XR) and generative AI, is paving the path toward more nuanced and personalized sensory experiences (Mourtzis et al., 2022; Wienrich & Latoschik, 2021). These technological advancements promise a future where the stimuli interpreted by the brain can be modified to create deeply curated experiences, considering all senses (Gao et al., 2021). In this promising future, the enhancement of individual enjoyment and engagement with various art forms and environments is not the sole outcome; it also holds significant potential for therapeutic and educational applications. Such a future entails the development of environments instilled to adapt and accommodate the unique sensory needs and preferences of every individual. This envisioned horizon represents a harmonious fusion of neuroscience, technology, and art, heralding an era characterized by unprecedented personalization in sensory experiences.

Innovations in immersive art experiences like the *House of Eternal Return* (2016) (https://www.youtube.com/watch?v=lmNvh0W6Y4Y) at the artist collective Meow Wolf in Santa Fe, New Mexico (Fig. 5.1) are at the forefront of transforming the sensory landscape, offering boundless, interactive, and multidimensional realms that respond to and are shaped by the visitor. In the *House of Eternal Return* there is an immersive adventure where each room unfolds into fantastical realms, urging visitors to touch, explore, and engage with the ever-shifting landscape. Interactive elements abound as visitors crawl through washing machines, enter surreal treehouses, or discover hidden portals within coat closets. Designed to tantalize the senses, actors and hosts enhance the tactile and visual experience by guiding guests through this vivid, touch-centric journey of discovery and wonder. Likewise, the international art collective known as teamLab housed in Tokyo, Japan (https://www.teamlab.art/) uses digital technology for immersive, experiential works. The immersive creations (Fig. 5.2) delve into the confluence of art, sensory perception, human corporeality, and technological innovation. Utilizing advanced methodologies such as projection mapping and motion capture, their immersive galleries stand as premier attractions in Tokyo, captivating a global audience with their pioneering approach. These spaces offer visitors an interactive experience where art and technology harmoniously intersect, engaging the body and senses.

Similarly, the intersection of technology and neuroarts is revolutionizing therapeutic applications. For instance, Ari Peralta, a prominent figure in sensory design and neuroscience, is pioneering in understanding how sensory triggers and textures influence human emotion and behavior, offering groundbreaking perspectives in creating sensory-optimized environments (Siamptani, 2022). In his research studio for sensory design, neuroaesthetics, and well-being, Arigami (https://www.arigami.co.uk/), the integration of nature-infused spaces and high-definition imagery in creating calming environments is another promising development. Another research lab at UCLA also seeks to understand the potential of arts in self-regulation and well-being. In their project Arts Impact Measurement system (https://soundhealth.ucsf.edu/content/ucla-arts-impact-measurement-system), psychometric tools are being used to understand the role of art in self-regulation and meditative states. The goal of such projects is to help identify beneficial aesthetic experiences for different populations and contexts, from elementary school children to first responders. As noted, the neuroarts installation *A Space for Being* in Milan, crafted by the Google hardware design team in 2019, stands as a notable exemplar. The exhibition elucidates the way distinct sensory environments provoke diverse physiological reactions, underscoring the imperative of being cognizant of one's physiological and aesthetic preferences. According to Ciangola (2023), the variety of installations empowers visitors to ascertain which environments most beneficially align with their health and well-being. This is achieved by providing insights into how their bodies and minds respond to varying artistic and sensory stimuli, accentuating the necessity of understanding and leveraging these responses to augment well-being and cognitive engagement.

The investigation of sensory enhancement and neuroaesthetic encounters within the realm of personalized artistic experiences is an emerging domain merging the most recent advancements in technology and neuroscience (Gruber, 2020). In the context of neuroaesthetics, sensory enhancement employs state-of-the-art technology to amplify or alter sensory interactions linked to art, culminating in heightened engagement and individualized artistic encounters. Recent advancements in this area focus on enhancing the sensory experiences of individuals through various

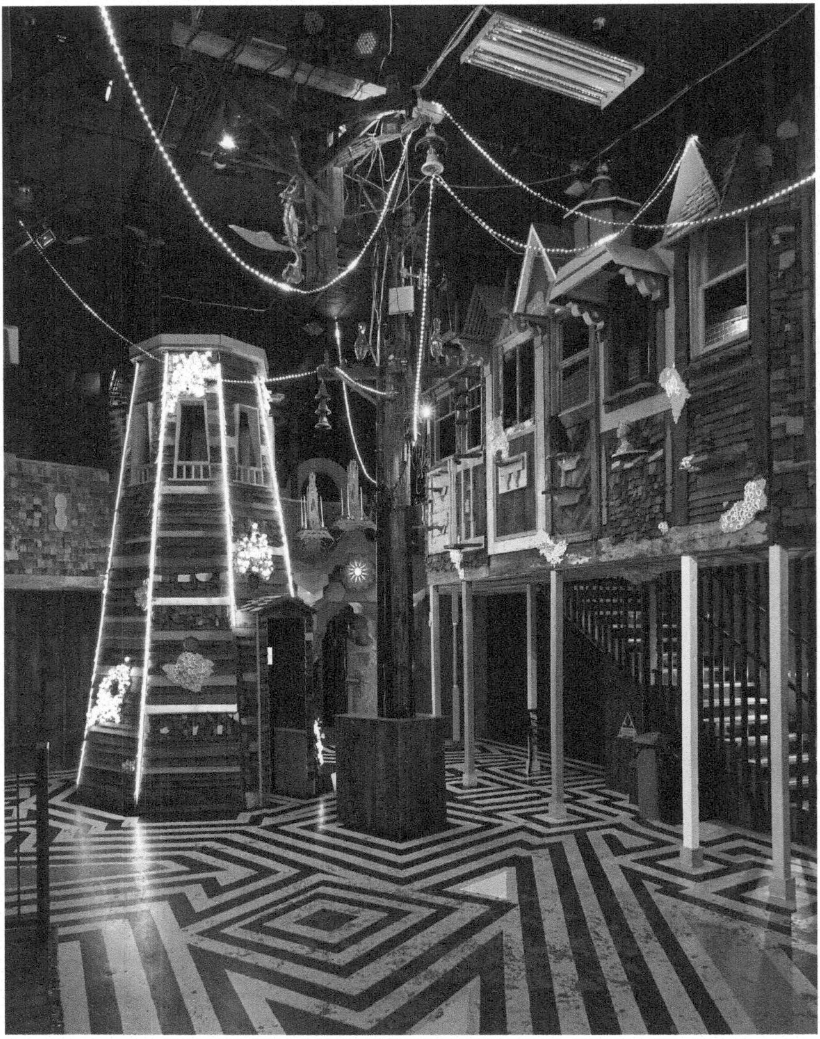

Fig. 5.1 Fancy Town, Meow Wolf, Santa Fe (*Source* From Wikimedia Commons, licensed under CC-BY 2)

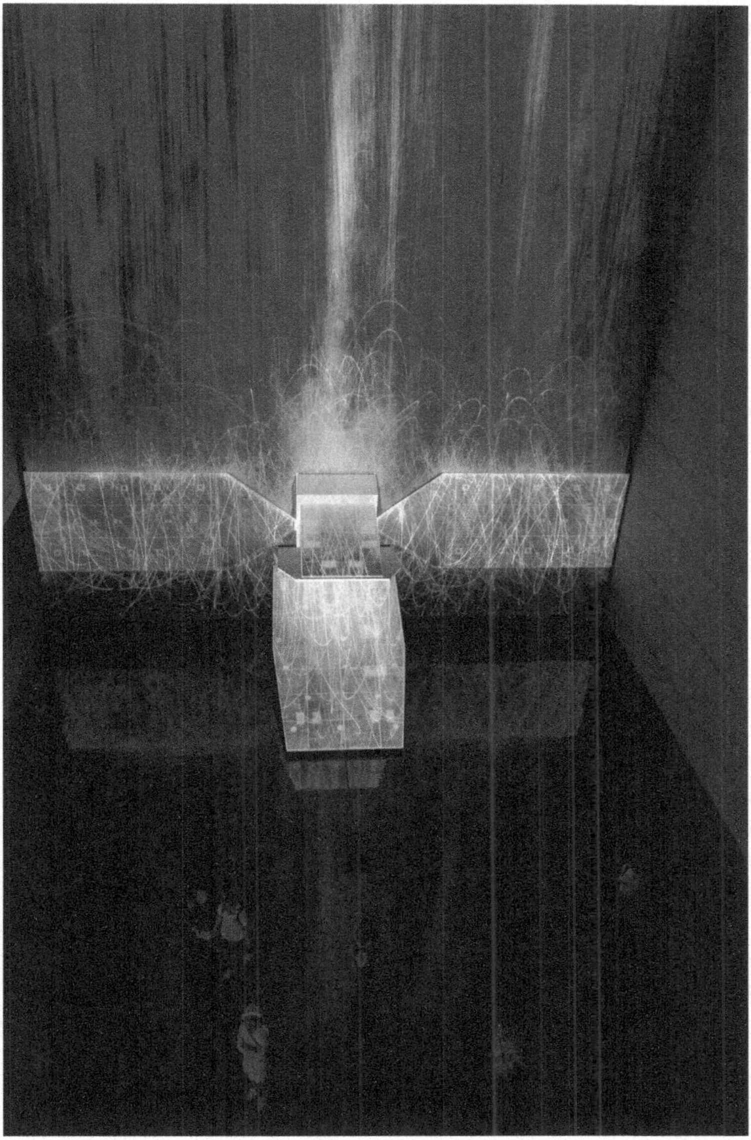

Fig. 5.2 Yoshikazu Takada, *Universe of Water Particles Under Satellite's Gravity*, 2014. TeamLab, Tokyo (*Source* From Wikimedia Commons, licensed under CC-BY 2)

means, including VR, AI-driven algorithms, and multisensory engagement (Grabbe, 2021; Lee, 2023; Magsamen & Ross, 2023; Ross & Snyder, 2023). Technologies like virtual reality create immersive art environments, while AI algorithms personalize experiences based on individual neurological feedback, intensifying the sensory experience and potentially fostering greater appreciation and understanding of art.

A subfield of particular relevance to this study is the domain of sensory augmentation for neuroaesthetic encounters, which is an emerging discipline. Conversely, auditory agnosia, characterized by the incapacity to identify or distinguish sounds, can impact both artists and viewers within the domain of auditory art, including music and sound installations. Individuals with auditory agnosia might rely more on visual cues or other sensory inputs to engage with auditory art, or they might interpret sounds in a fundamentally different way, leading to unique creations or experiences of auditory art sensory substitution, addition, and expansion via various devices and techniques (Consorti et al., 2023). Recent research has focused on the use of haptic and electro-tactile devices to create new sensorimotor contingencies, leading to sensory addition and expansion rather than substitution. These approaches have the potential to enhance aesthetic experiences and may have therapeutic applications (Eagleman & Perrotta, 2023; Kourtesis et al., 2022). Studies have also explored the use of sensory augmentation devices to extend visual-spatial perception, indicating the possibility of shaping and reshaping the aesthetic brain (Shvadron et al., 2023). Advancements also include sensory substitution devices like tactile-vision and feelSpace belts, which replace or augment senses using another sense, such as vibrotactile stimulation. This allows users to perceive their environment in novel ways and enrich their interaction with art (Macpherson, 2018). Artificial sensory memory devices, incorporating bioinspired sensing and neuromorphic engineering technologies, are being developed for recognition, manipulation, and learning applications (Wan et al., 2020). These devices expand the boundaries of how we perceive and interact with art and leverage it for therapeutic purposes.

The latest research and developments in virtual art for therapy and interactive art experiences show promising results for their use in mental health and well-being. Several studies and articles highlight the positive impact of virtual art on mood, anxiety, and mental health (Chan et al., 2023; Cotter et al., 2023; Pira et al., 2023; Ruívo et al., 2023). VR and collaborative virtual environments (CVE) are being explored as effective mediums for arts therapy, offering novel opportunities for artistic expression, self-improvement, and motivation for psychotherapy (Li & Yip, 2023). Furthermore, the incorporation of virtual reality (VR) into the realm of art therapy has been examined as a means of providing therapeutic assistance to individuals, including adolescents (Hutson, 2022). There is a growing body of evidence suggesting that participating in sensory and tactile endeavors, such as art therapy, may have the potential to enhance brain plasticity, alleviate anxiety, and foster connections and optimism (McGorrill, 2023).

Exploration of multisensory environments and tactile modalities within the domain of neuroaesthetics research is currently underway, aimed at comprehending their influence on perceptions, emotions, and neural responses. These investigations offer fresh insights into the development of sensory-rich experiences. Concurrently, recent research endeavors and initiatives focused on somatic interactions in artistic encounters underscore the profound impact of the body on creativity and aesthetic experiences (Stepanova et al., 2023). Somatic experiences, which establish a connection between bodily awareness and psychological phenomena, are being investigated within the context of art therapy and trauma recovery. Here, the comprehension of physical sensations and the mind–body connection holds paramount significance in the healing process (Schlief, 2023). Furthermore, the field of somaesthetics, which explores how artists utilize their bodies in the creation of artworks and how the body assumes a formative role in diverse art forms, is gaining prominence (Shusterman & Veres, 2023).

Moreover, the emerging field of computational neuroaesthetics seeks to bridge the divide between neuroaesthetics and computer science, presenting promising methodologies for quantifying and simulating brain activities associated with aesthetic encounters (Zeng et al., 2023). Simultaneously, the realm of sensory experiences enhanced by AI is witnessing a growing interest in unraveling the biological underpinnings of aesthetic perceptions and harnessing this knowledge to craft more captivating and impactful sensory environments (Andres et al., 2023). This interdisciplinary approach holds the potential to inform the development of sensory-rich experiences that resonate with the human psyche and evoke positive emotional responses. This interdisciplinary approach has the potential to shape the design of sensory-rich experiences that deeply connect with the human psyche and elicit positive emotional responses.

The integration of AI and neuroaesthetics is paving the way for the development of sensory environments that go beyond visual appeal, offering cognitive and emotional enrichment. However, it's crucial to acknowledge that the success of these cutting-edge technological interventions hinges on their alignment with the unique characteristics and preferences of the intended neurotypes.

The burgeoning landscape of neuroaesthetic technology, with its promise of personalized sensory experiences, necessitates the development of taxonomic criteria to categorize sense experience by neurotribe. As we delve deeper into the realm of augmented and virtual sensory environments, the precision in understanding and categorizing individual sensory profiles becomes paramount. Without a clear, systematic approach to classify how different neurotribes experience and interpret sensory stimuli, the full potential of these technologies cannot be realized. This is particularly important given the subjective nature of sensory experiences and the wide variance across individuals, especially within neurodivergent communities. Developing an assistive diagnostic tool able to accurately identify and categorize individual aesthetic preferences and sensory responses will enable advanced technologies to be utilized effectively, leading to truly customized and beneficial sensory experiences.

In the realm of sensory experience and neuroaesthetics, the metaphor of the video game *Rez* (2001, 2023) (https://youtu.be/Ycc-C6QP16k?si=f4Km_gdjxKWlhwA2) provides a compelling parallel. Just as players in *Rez* find that as they progress and improve, the audiovisual landscape of the game becomes increasingly rich and immersive, individuals can similarly curate their sensory environments to better suit their needs, enhancing comfort, focus, or pleasure. The game itself, a musical rail shooter known for its synesthetic blend of sound, visual, and tactile feedback, offers an illustrative example of how sensory experiences can be tailored and intensified through practice and understanding. As players navigate through the abstract cyberworld, the gameplay and its impacts are synchronized with the music, creating a layered sensory experience that mirrors the journey of life and technological singularity. The evolution of *Rez*, from its initial release on Dreamcast and PlayStation 2 to its expansion into virtual reality platforms, underscores the continuous advancement in sensory augmentation technologies. Similarly, as individuals better understand and cater to their unique sensory profiles, particularly in the context of neurodiversity, they can leverage these technological advancements to enrich their daily experiences. The development and application of taxonomic criteria and assistive diagnostic tools in neuroaesthetics are crucial in this regard, promising more empathetic and effective interventions that elevate the quality of life and artistic engagement.

Developing an application to assist individuals in discovering their sensory profiles and aesthetic preferences represents a significant advancement in personalized neuroaesthetic experiences (Table 5.1). Such an application would ideally balance instinctual sensory quizzes with personalized, adaptive quizzes to understand and grow with the user's sensory experiences and preferences. It would explore reduced chromoscapes to accommodate unique visual profiles and continuously learn from the user's diary entries about their art interactions, thereby refining its recommendations.

This application would create a detailed sensory profile for each user, considering their individual reactions to various stimuli like touch, smell, sound, and visuals. It would integrate considerations for neurodiversity, recognizing the commonalities and specific needs of users with conditions such as sensory processing disorder. By providing a platform for users to track and reflect on their sensory experiences, the app would facilitate a deeper understanding of personal aesthetic preferences and how these interact with one's neurotype. A key feature of this app would be its ability to curate personalized "neurosafe spaces"—virtual or real environments where users can engage with arts and sensory experiences tailored to their profiles. This could extend to recommending art exhibitions, immersive installations, or even natural settings that resonate with user sensory preferences. The app would also foster community by connecting users with similar sensory profiles, encouraging the exchange of experiences and recommendations.

For those interested in creative activities, the app would map out sensory profiles to suitable crafts or hands-on art techniques, recommending activities aligned with user sensory and skill preferences. It would leverage AI to analyze oeuvres of artists, helping users discover artworks and styles that resonate with their sensory profile. Furthermore, the app would use AI to gauge color preferences and potential visual spectrum alignments, such as color blindness, enhancing the personalization of the art experience. The app would also cater to a diverse range of users by providing options to modify sensory experiences according to individual needs, whether it's adjusting the light in an art exhibit or the sound

Table 5.1 Criteria for sensory profile quizzes

Consideration category	*Specific aspects*	*Description/notes*
Quiz type	Instinctual vs. personalized quizzes	Balance instinctual sensory quizzes with personalized, adaptive quizzes that learn from the user's interactions and diary entries about their art experiences
Sensory area	Reduced chromoscapes	Explore reduced chromoscapes and accommodate unique visual profiles, particularly for individuals with color vision deficiencies or preferences
Research path	Future research directions	Encourage users to keep a diary of their reactions to different types of art, note their sensory triggers, and analyze patterns over time
Application features	Creation of sensory profile	Develop detailed sensory profiles considering reactions to touch, smell, sound, visuals, and more
Community aspect	Finding similar profiles	Include a feature to find others with similar sensory profiles nearby, fostering a community of shared experiences and recommendations
Personalization	Tracking and habit identification	Use the app to track habits and sensory reactions to help identify the user's neurotype and refine recommendations
Art engagement	Pre-visit profile and artistic discovery	Allow users to tag favorites and discover new artistic preferences. Use pre-visit profiles to curate virtual safe spaces or recommend physical art spaces
Sensory exploration	Textural and color preferences	Ask users about their texture and color preferences, mapping these to potential crafts or hands-on art techniques
Artistic aid	AI for artist oeuvre and color preferences analysis	Use AI to analyze artists' works to distinguish color preferences, potential for color blindness, or other visual spectrum alignments
Neuro-needs alignment	Bespoke conversation and interaction patterns	Tailor conversation patterns and app interactions to align with the user's neuro-needs, making the experience as comfortable and engaging as possible

in a musical piece. It would aim to create bespoke conversation patterns and interactive experiences that align with the user's unique neurological needs, offering a comfortable and understanding user interface.

In essence, this application would represent a convergence of technology, art, and personal well-being, offering a tool for individuals to explore and engage with their sensory world more deeply and meaningfully. By considering the nuances of personal experience and the diverse spectrum of neurotypes, the app would facilitate a more empathetic and effective deployment of neuroaesthetic interventions, enhancing both artistic enjoyment and therapeutic outcomes. With such technology, individuals could navigate their sensory environments as skillfully as players navigate the evolving audiovisual world of the video game *Rez*, optimizing their experiences for comfort, engagement, and enjoyment.

5.2 Embodied Neuroaesthetics: Tailored Sensory Experiences

Contemplating the future applications of an app dedicated to embodied neuroaesthetics and tailored sensory experiences envisions a seamless fusion of technology and individual sensory comprehension. This AI-driven virtual "cabinet of curiosities," reminiscent of a modern Wunderkammer, holds the potential to become a novel arena for artistic exploration. Within its digital confines, each neuroaesthetic profile can guide users through a labyrinth of art that uniquely resonates with their senses. This innovation not only proposes to revolutionize our interactions with art, but also to expand our understanding of how we perceive and interpret sensory information. In an era marked by the ongoing exploration and expansion of the boundaries of sensory experiences, this application stands as a testament to human ingenuity. It offers a personalized journey into the realms of embodied neuroaesthetics, setting the stage for a future where each individual's sensory world is as vividly understood as it is uniquely experienced. Moreover, the technology required to facilitate such personalized experiences is now widely accessible to the general public, ushering in a more inclusive and resonant era in sensory engagement with art.

By enhancing sight, sound, touch, and even taste, all readers of this book now possess the means to customize their perception of the world. The accessibility of augmented reality (AR) smart glasses and contact lenses is on the rise, providing the capability to adjust or enhance visual stimuli in real-time. The initial setback experienced with the launch of Google Glass in 2013, coupled with the widespread acceptance of smartphones across diverse demographic groups, has mitigated earlier privacy concerns. This has paved the way for the emergence of a new generation of devices, as noted by Zuraikat in 2020. As of 2024, several options are available to connect via Bluetooth to one's smartphone or device to get directions, identify what is being looked at, or to overlay various elements. These include Ray-Ban Stories and Razer Anzu for general use, NuEyes Pro 3 for accessibility and assistive technology, and Bose Frames Tempo for music and entertainment (England et al., 2023; Pfeifer et al., 2023). There are even AR contact lenses like Mojo Vision (Kazanskiy et al., 2023). These devices can overlay digital information onto the physical world, altering or augmenting what individuals perceive, tailored to their unique sensory profile. For example, someone sensitive to bright lights might use AR to dim their perception of overly bright environments or highlight certain colors that are aesthetically pleasing and calming to them or to ameliorate colorblindness (Ananto et al., 2011; Tillem & Gün, 2023).

The landscape of modern hearing technologies is moving further into personalization (Glista et al., 2023). For auditory experiences, personalized wireless earbuds such as PerL adapt to the user's unique hearing profile, modulating auditory input to suit individual preferences and needs. A range of personalized options are now available that connect via Bluetooth to any device, including nura Nuraphone headphones, while others are specifically tailored for hearing impairment like the Avantree Ensemble, SIMOLIO, and even BenexBesdi bone conduction headphones (Toya et al., 2023). This technology can improve auditory signals for those with difficulties amplify or soften sounds, filtering out background noise, or even adjusting the audio quality of music or spoken words, making auditory experiences more enjoyable and less overwhelming for those with sensitive hearing.

The domain of haptic technology has made significant strides in replicating tactile sensations, thereby enhancing user interaction and experience in various fields. Haptics refer to the use of technology that stimulates the senses of touch and motion, primarily to recreate the experience of interaction with virtual objects (Tong et al., 2023). One of the most compelling advancements in this field is the development of haptic gloves and suits, which allow users to feel and manipulate virtual objects as if they were real, offering applications ranging from virtual reality gaming to remote surgery (Feick et al., 2023). For instance, there are a variety of haptic gloves that range from simulating pressure with the SenseGlove to temperature and texture with WEART and all the way to a full system with air compressor and backpack with the HaptX Gloves G1 system (Shen et al., 2023). In the realm of virtual and augmented reality, haptics enhance the immersive experience by allowing users to feel the weight, texture, and temperature of virtual objects, significantly bridging the gap between virtual and physical worlds.

In the realm of olfactory experiences, technologies like OVR (Olfactory Virtual Reality) can release scents correlated with specific locations or events, creating a multisensory environment. Software plugins can be easily adapted for digital and virtual content for game engines like Unity or Unreal. The simple plug-in attaches to most head-mounted displays (HMD) or virtual reality headsets and release scents as programmed. This can be particularly beneficial for individuals whose mood or cognitive function is influenced by certain smells, allowing for a controlled and personalized scent experience amplifying their overall sensory environment (Zhong et al., 2023).

Finally, VR technology has the potential to transport individuals to entirely new locations, combining visual, auditory, and even olfactory and tactile technologies for a fully immersive experience. The latest iterations of hardware, including the Meta Quest 3 and Apple Vision Pro, further leverage the combination of both VR and mixed reality (MR) technologies to move seamlessly between them (Waisberg et al., 2023). When customized according to individual sensory profiles and neurotypes, VR has the potential to offer therapeutic, educational, or purely recreational experiences that are not only captivating but also comfortable and enjoyable for users (Liu et al., 2023). Collectively, these technologies mark a new frontier in creating immersive environments crafted around individual sensory profiles, pointing toward a future where everyone can embrace and interact with their surroundings in a manner that aligns with their comfort and satisfaction.

In selecting technologies to accommodate various sensory profiles, the salutogenic model provides a critical framework. The basis of Salutogenesis in the context of sensory adaptation and aesthetic preferences is to foster environments and interactions that enhance individual sense of well-being, health, and ability to navigate their surroundings effectively (Mittelmark et al., 2022). The approach, which is rooted in the origins of health and well-being, contrasts the traditional pathogenic focus on disease and dysfunction (Bauer et al., 2020). In fact, the term "salutogenesis" derives from the Latin word *salus*, meaning health, and the Greek word *genesis*, meaning origin. The salutogenic model was initially developed through studies of Holocaust survivors, focusing on understanding how individuals cope with stress and maintain health even in the face of extreme adversity (Reed & Haugan, 2021). In the context here, the model can assist in providing insights to frame interventions target embodiment.

Embodiment, or the deep, often subconscious, perception of inhabiting a body, is rooted in sensorimotor processes. Technologies that strengthen this innate bodily awareness can significantly enhance orientation and interaction with physical space, thereby enhancing the first-person perspective and overall sensory experience (Lynch & Del Casino, 2021). This is particularly relevant in creating tailored sensory experiences, as a stronger sense of embodiment can lead to more profound and enjoyable interactions with art and the environment. The experience of deep emotion and the perception of being grounded and connected can often diminish due to various factors like trauma, brain damage, or PTSD (Lyon et al., 2021). Traumatogenic dissociative symptomology manifests as severe distortions in how the bodily self is perceived in relation to the outside world, often corresponding with trauma during critical periods of sensorimotor development (Bloomfield et al., 2020; White et al., 2022). Therefore, technology that reconnects the brain and body to reinstate states of safety is vital for a return to the foundation of selfhood. This can be accomplished through somatic sensory-based therapeutic methods. EEG neurofeedback training, a non-invasive approach, is predicated on the concept that cognitive and motor functions rely on effective communication among neuronal assemblies organized into networks (Viviani & Vallesi, 2021). By monitoring real-time neuronal activity, this method has the potential to enhance high-amplitude alpha oscillations linked to a tranquil and introspective mental state (Karvat et al., 2020). In essence, it can assist individuals in attaining a more balanced and peaceful state of mind.

Expressive arts therapies offer another avenue for the integration of technology. Each modality of expressive arts therapy incorporates physical movement and purposeful actions, leading to deliberate creative expressions. These engagements establish or reestablish sensory-motor feedback loops, fostering a sense of empowerment, agency, and a positive connection with one's surroundings. Furthermore, exposure to enjoyable environments, potentially enhanced or generated with the assistance of technology, can raise the threshold for amygdala activation, facilitating emotional resilience and reducing reactivity. A salutogenic approach in technology and arts can lead to the development of interventions and experiences that not only cater to individual sensory and aesthetic preferences but also promote a profound sense of health, well-being, and embodiment. This approach underscores the importance of considering the whole person—their neurological, emotional, and physical states—in creating art and therapeutic experiences that resonate on a deeper, more personal level.

Past salutogenic interventions have utilized advanced technology to elicit particular sensory encounters. Nevertheless, there exists a historical antecedent that offers a distinct paradigm for the present discourse on the convergence of the arts and the cultivation of emotional responses. Prior to the emergence of contemporary national museums across Europe during the nineteenth century, artworks were primarily displayed in religious establishments such as churches or exclusively held in private collections accessible only to the privileged elite. These early collections, including *Wunderkammer* and *Kunstkammer*, with their eclectic and extensive collections, symbolize the human desire to understand and connect with the world. The renaissance-era *Wunderkammer*, literally translated from the German as "room of wonder," or more commonly in English as "cabinet of curiosities," is historically significant as a collection of rare, exotic, or intriguing objects displayed in a single room or cabinet. These collections, which often encompassed natural history, art, and antiquities, were a precursor to modern museums, becoming popular among European nobility and wealthy collectors (Williams & Tsien, 2013). They were designed for private viewing in either an actual cabinet (Fig. 5.3) or a single room (Fig. 5.4), showcasing the sentiment of wonder and the breadth of human knowledge and creativity. Similarly, the *Kunstkammer* (Fig. 5.5), meaning "art chamber," is akin to a *Wunderkammer* but with a specific focus on art collections. This includes fine arts, sculptures, and often valuable, rare, or unusual decorative objects. These collections represent an early form of the art museum and were typically, like *Wunderkammer*, assembled by European nobility or wealthy collectors. Unlike the curiosity cabinets, however, which was often confined to a single room, *Kunstkammer* collections were usually displayed in large halls, showcasing large works of art (Pilaski & Kaliardos, 2013).

For the purposes of this study, the *Wunderkammer* is metaphorically likened to creating experiences that evoke atmospheric and environmental reactions, while the *Kunstkammer* symbolizes the curated display of a wide variety of art and creative expressions. These historical collections

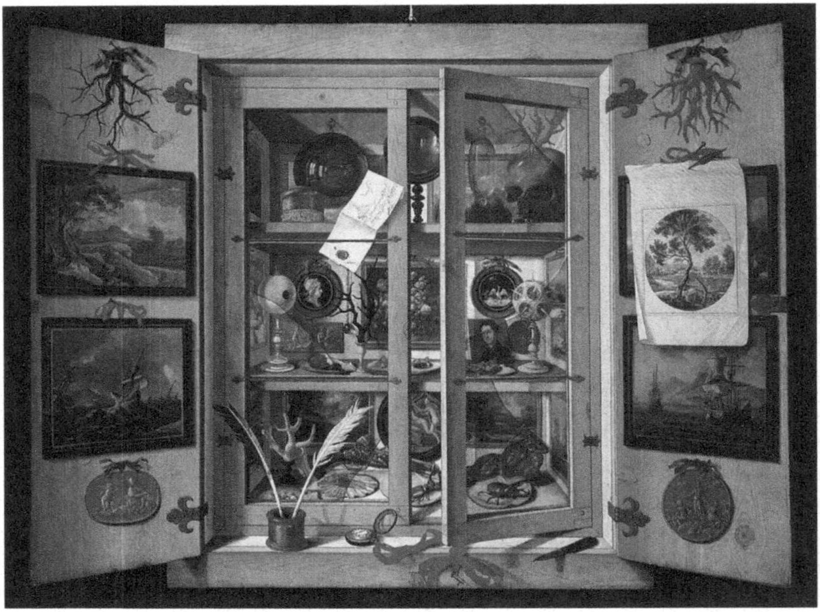

Fig. 5.3 Domenico Remps, *Cabinet of Curiosities*, 1690 (*Source* From Wikimedia Commons, licensed under CC0)

inspire the use of modern technology to craft immersive, personalized sensory experiences, aligning with the salutogenic model that focuses on promoting health and well-being through art. The latest technological advancements enable the creation of virtual spaces tailored to individual sensory profiles and aesthetic preferences, reminiscent of the personal reflection and wonder once found in these rooms of wonder. This modern approach continues the age-old human pursuit of knowledge, beauty, and emotional connection through art, enriching it with the capabilities of current technology.

A virtual *Wunderkammer* thus emerges as a novel concept where the essence of a traditional cabinet of curiosities as place of wonder is transformed into a digital space, expanded, and personalized through the XR, haptic, auditory, and olfactory technologies mentioned and created and curated via generative artificial intelligence (GAI). The digital space becomes a dynamic and responsive gallery, adapting and evolving

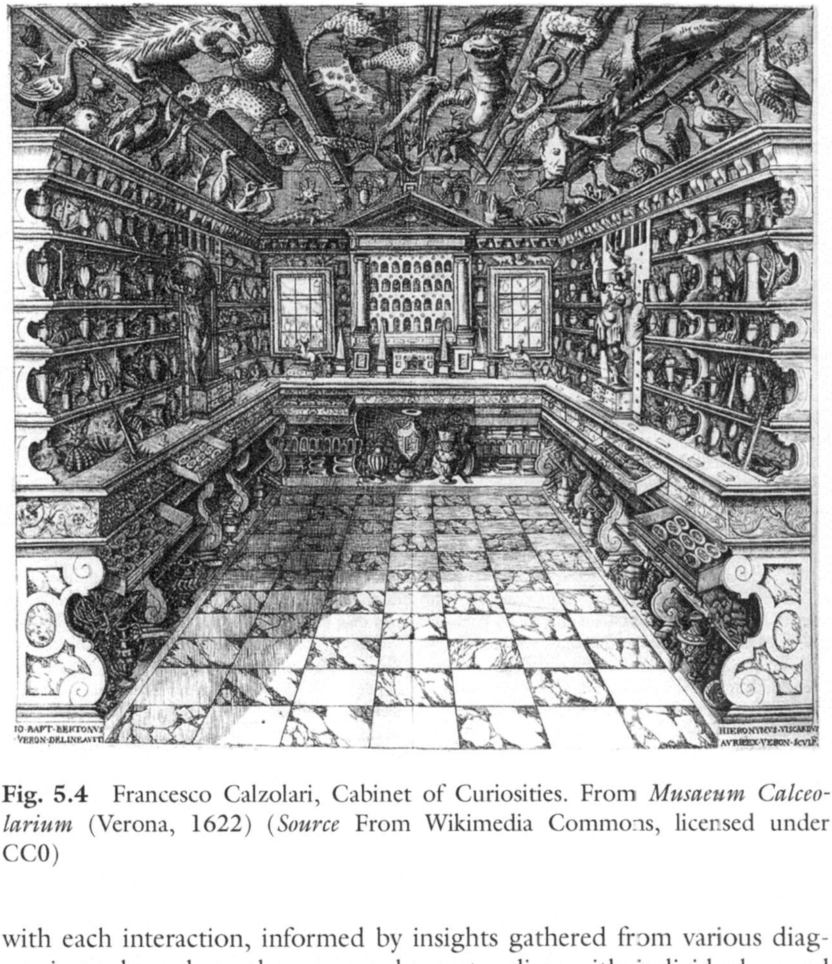

Fig. 5.4 Francesco Calzolari, Cabinet of Curiosities. From *Musaeum Calceolarium* (Verona, 1622) (*Source* From Wikimedia Commons, licensed under CC0)

with each interaction, informed by insights gathered from various diagnostic tools such as the proposed app to align with individual neural profiles. Notably, several examples of readymade immersive experiences already exist. Digital wellness platforms like *TRIPP* and *VR Zen* for the Meta Quest and Vuvana Immersive Serenity for Android offer guided or unguided meditations, along with deep breathing. *MOOD* is another VR app that offers lessons from mental health practitioners to help achieve peace of mind. There are even VR experiences that replicate neurodivergent experiences like that of schizophrenia. *Goliath: Playing with*

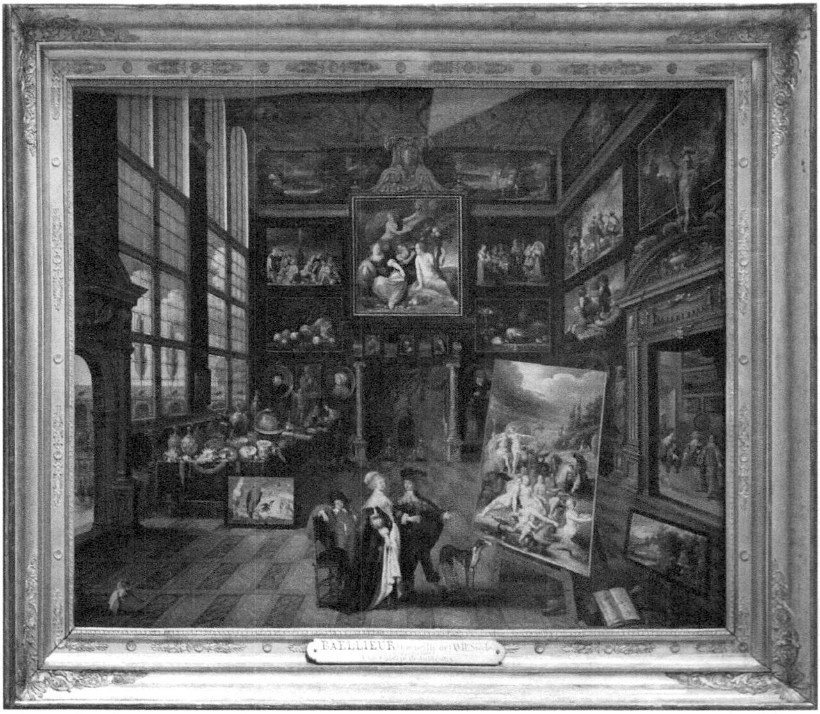

Fig. 5.5 Hans Jordaens III, *Kunstkammer*, 1637 (*Source* From Wikimedia Commons, licensed under CC-BY 4)

Reality (https://goliathvr.io/) (2021), for instance, is a twenty-five-minute animated VR experience that brings users into the disorienting world of the condition. Still other apps allow users to create art in a variety of media from sculpture with *SculptrVR* to painting with *Painting VR*, and even pottery with *Pottery VR*. The limitation of each of these examples, however, is that the experience is not changeable.

It is important to note the continual progress beyond these simulations and moving beyond predetermined reactions from users toward adaptive and personalized ones which evolve in accordance with users' requirements. This advancement has been facilitated by the developments in Generative Artificial Intelligence (GAI) in 2023. The general public was first introduced to AI-generated art through social media applications in the summer of 2022, with platforms like Lensa enabling users to generate avatars and render images. In February 2023, the stable release of ChatGPT 3 witnessed the most rapid adoption of any technology in history, amassing a million users in under one month. Such large language models (LLM) are becoming increasingly sophisticated and can generate text, code, and now imagery with multimodal AI. With these, users can easily and freely generate any text or image desired, but summer of 2023 also saw the development of text-to-video tools. Now, there are many to choose from with Synthesia, InVideo AI, Deepbrain AI, and more creating cinematics for the general public.

The progress in GAI is exemplified by Stability AI's SDXL Turbo, which enables real-time text-to-image generation. These advancements have unlocked unprecedented possibilities in content creation, democratizing the process and allowing even individuals without specialized expertise to craft intricate and nuanced artwork. In essence, this means that anyone now has the capability to generate diverse content, whereas previously, creating a VR simulation required a team of specialists, including 3D modelers, coders, UX designers, game designers, and more. This democratization of creative tools expands the horizons of personalized art experiences, granting individuals the power to explore and express their unique sensory and emotional landscapes.

With the advent of sophisticated AI technologies and a detailed understanding of individual sensory profiles, there is an emerging potential to not just create, but also curate personalized art experiences.

AI algorithms, equipped with user-specific data encompassing sensory preferences, neurotype, and desired emotional outcomes, have the potential to serve as discerning curators within virtual domains. These intelligent systems would meticulously sift through extensive repositories of art, music, and sensory stimuli, meticulously assembling a unique collection of experiences tailored precisely to the individual's tastes and requirements. For instance, an individual seeking a calming and immersive encounter might encounter a curated playlist featuring soft, ambient soundscapes, visually soothing artworks, and gentle tactile feedback. These elements would be thoughtfully synchronized to induce a state of relaxation. This AI-driven curation process ensures that every interaction within the virtual environment not only captivates and delights but also deeply resonates and offers tangible benefits, aligning with the individual's neuro-needs. This degree of personalization allows experiences which are no longer passive consumptions but actively honed by users themselves, facilitated by the intelligent mediation of AI.

At the same time, technology cannot do everything itself. As one interviewee for this study insightfully noted, "AI can only draw from what humans have already done." Therefore, the essence and depth of human experience are crucial in guiding AI to create resonant and meaningful experiences and require a human-in-the-loop strategy. This human insight is particularly vital when considering the vast spectrum of neurodiversity and individualized sensory processing. AI, while potent in its capability to generate and tailor experiences, relies on the rich tapestry of human emotions, sensations, and creative expressions to truly capture and enhance the wonder of art. Thus, an AI-augmented virtual Wunderkammer, shaped by human ingenuity and sensory profiles, stands as the forthcoming realm of art engagement and experience. It presents the prospect of profoundly personalized, immersive, and emotionally evocative encounters, marking the dawn of the exploration of neural and physiological states associated with wonder.

In an illustrative scenario, drawn from the insights of synesthesia interviews, imagine an individual entering this virtual space, equipped with AR glasses and haptic feedback devices. As they navigate through the simulation, the AI, having learned their preferences, unveils a curated selection of artworks, each resonating with their unique sensory and emotional constitution. A piece of music, visually represented, shifts and changes color in harmony with their mood, while the scent of a distant memory is subtly released, enhancing the emotional depth of the moment. Alternatively, in considering ideal spaces, one could create a safe familiar space with elements that are desired, such as diffused lighting, bird sounds from an open window, and plants (Fig. 5.6), or remembering a first date with one's spouse being by a duckpond on a warm spring day (Fig. 5.7). These encounters extend beyond mere visual and auditory elements, embracing a harmonious fusion of sensations meticulously customized to each individual's neurological and emotional constitution. This vision encapsulates the transformative possibilities inherent in such a virtual therapeutic milieu—a realm where technology, art, and human essence harmonize to forge an entirely novel dimension of aesthetic engagement. Nonetheless, it is crucial to address ethical considerations regarding the collection of user information and content generation to facilitate these experiences.

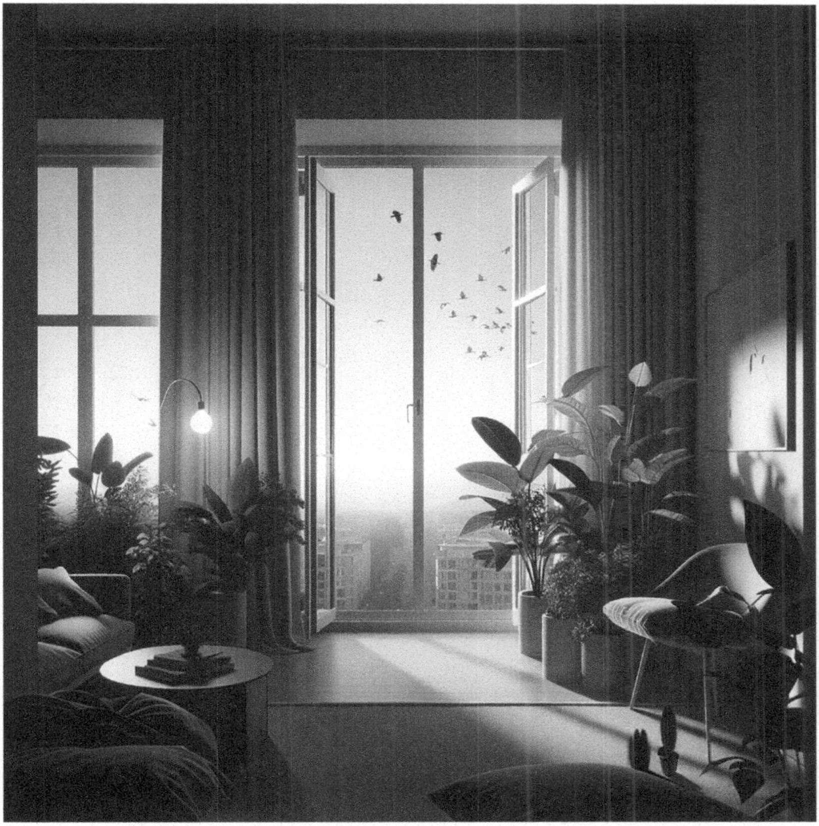

Fig. 5.6 Cozy Interior Space with Open Window and Birds, 2024 (*Source* Permission by authors)

In the context of embracing and integrating AI and neuroarts to craft personalized sensory experiences, several ethical considerations must be addressed to ensure these innovations benefit individuals and society responsibly. The development and deployment of neuroaesthetic experiences demand the establishment of a robust ethical framework, one that delves into the profound implications of manipulating sensory encounters and staunchly ensures these transformative technologies are employed to elevate well-being rather than exploit vulnerabilities. Within this ethical landscape, comprehensive guidelines must be put in place to govern

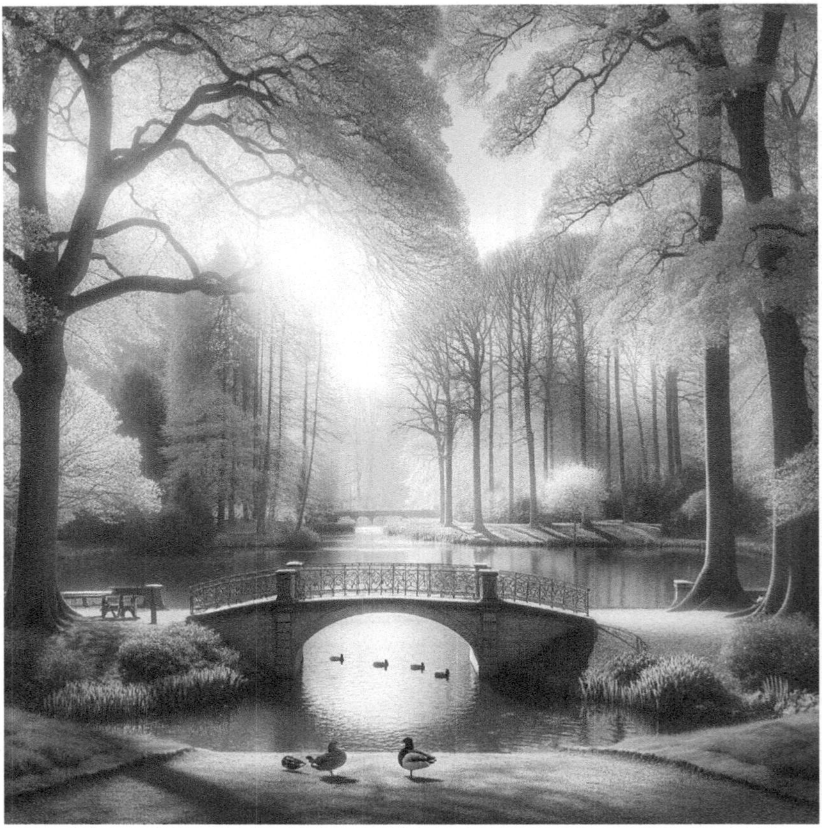

Fig. 5.7 Afternoon at a Duck Pond, 2024 (*Source* Permission by authors)

the entire lifecycle of neuroartistic content, encompassing its creation, curation, and distribution. These guidelines must steadfastly uphold the principles of respecting cultural diversity, safeguarding individual dignity, and staunchly refraining from perpetuating stereotypes or biases. In the sphere of neuroarts, personal data transcends mere preference history; it extends to include sensitive neurological and physiological information. Ensuring informed consent becomes paramount, requiring that users possess full awareness of the data collection processes, the purposes to which their data will be applied, and the potential risks involved. To this end, users should retain uncompromising control over their data, with transparent options for opting in or out of data collection, along with the ability to access, rectify, or delete their information at their discretion. Privacy safeguards should be stringent, rigorously assuring that data remains anonymized, securely stored, and shared only with the explicit consent of the individuals to whom it pertains.

As neuroarts and personalized sensory experiences become more prevalent, a pertinent concern arises regarding the potential for commercialization to pave the way for the commodification of personal experiences and cultural expressions. This risk looms large, potentially leading to a scenario in which art and sensory encounters become homogenized, shaped primarily by market trends rather than authentic personalization. Thus, it becomes imperative to cultivate an ecosystem that strikes a delicate balance between innovation and the preservation of culture and diversity. Therefore, a concerted and deliberate effort is warranted to ensure that these groundbreaking technologies remain accessible to all, transcending economic boundaries and preventing the emergence of a societal divide in which only a privileged subset can reap the benefits of these advancements. By meticulously addressing these ethical considerations, we can guide the trajectory of neuroarts and AI-driven personalization toward a future that not only upholds individual autonomy but also champions inclusivity. By addressing these ethical considerations, we can guide the development of neuroarts and AI-driven personalization toward a future that respects individual autonomy, promotes inclusiveness, and enriches the human experience by nurturing tailored sensory engagement which resonate with the diverse tapestry of humanity.

5.3 Final Considerations and Paths Forward

In the intersection of neuroarts and personalized sensory experiences, their impact on public engagement, education, healthcare, and wellness assumes growing significance. The integration of art education and neuroscientific literacy becomes imperative, facilitating an understanding of the profound influence of art on neural and physiological processes. This knowledge not only enhances art appreciation but also empowers individuals to leverage these insights for cognitive enrichment and personal well-being. Collaboration among educators, artists, and neuroscientists is essential in the development of interdisciplinary curricula and programs, fostering a society that recognizes and comprehends the intricate relationship between art and neural mechanisms. In the healthcare and wellness domains, neuroarts offers promising avenues for therapeutic interventions and stress reduction strategies. The utilization of personalized art experiences can lead to innovative non-pharmacological treatment options, ranging from the management of chronic pain to supporting the recovery process in various neurological conditions. In wellness settings, the integration of neuroart installations can provide individuals with spaces for meditation, stress relief, and emotional well-being.

Looking ahead, the future of neuroarts is shaped by continuous research, interdisciplinary collaboration, and accessibility for all. Ethical governance is essential for responsible development. As we enter the Fifth Industrial Revolution, technological progress accelerates, with brain–computer interfaces (BCIs) emerging as a significant advancement. BCIs enable direct brain-device communication, transforming how we interact with art and offering therapeutic potential for neurological disorders. This shift toward interactive AI and BCIs redefines art creation and enhances therapeutic possibilities. In other words, we will not need to type on a device to create content on another application or platform, but instead merely "think" what we want our devices to do. This technological marvel not only unlocks fresh ways for artistry and sensory exploration, it also harbors the potential to revolutionize therapy by offering innovative solutions for addressing neurological disorders and facilitating brain recovery.

When we peer into the future of neuroarts, we encounter a landscape filled with both challenges and opportunities. Ethical considerations, particularly when delving into the intricacies of human cognition and emotion, loom large. Privacy, consent, and the risk of manipulation demand careful navigation. Equally vital is the challenge of ensuring that these advancements benefit all strata of society, rather than a privileged few. Yet, amidst these challenges, a world of possibilities awaits. Neuroarts has the power to deepen our comprehension of human experiences, broaden the horizons of creativity, and enhance mental well-being and cognitive prowess. The fusion of BCIs and emerging technologies may usher in entirely novel art forms and ways of perceiving reality, pushing the frontiers of creativity and expression to unimaginable realms.

Forging the path for future research and practice in the captivating realm of neuroarts demands a dynamic, interdisciplinary approach that recognizes the profound intersection of neurology and art. Here, the realms of science and creativity converge, presenting us with a unique fusion of human cognition and artistic expression. Scientists, artists, ethicists, and technologists must join forces to navigate the complex ethical terrain and fully unlock the potential of these technologies. Continued research is our compass, guiding us to understand the deep connections between how these experiences influence our brains, and the ways they can be harnessed to enrich education, therapy, and recreation. As we embark on this adventure into the uncharted technological frontiers of neuroarts, we must do so with a potent blend of caution, curiosity, and an unwavering commitment to the fusion of neurology and art. It is this synergy which promises to enrich the human experience, promising a future where neuroarts not only captivate and inspire but also significantly enhance the health and well-being of individuals and society at large. In embracing both the challenges and opportunities which lie ahead, we can chart a course toward a brighter, more enriched future.

References

Ananto, B. S., Sari, R. F., & Harwahyu, R. (2011, November). Color transformation for color blind compensation on augmented reality system. In *2011 International Conference on User Science and Engineering (i-USEr)* (pp. 129–134). IEEE.

Andres, J., Ocampo, R., Bown, O., Hill, C., Pegram, C., Schmidt, A., Shave, J., & Wright, B. (2023, March). The human-built environment-natural environment relation-an immersive multisensory exploration with 'system of a sound'. In *Companion proceedings of the 28th international conference on intelligent user interfaces* (pp. 8–11).

Bauer, G. F., Roy, M., Bakibinga, P., Contu, P., Downe, S., Eriksson, M., Espnes, G. A., Jensen, B. B., Juvinya Canal, D., Lindström, B., Mana, A., Mittelmark, M. B., Morgan, A. R., Pelikan, J. M., Saboga-Nunes, L., Sagy, S., Shorey, S., Vaandrager, L., & Vinje, H. F. (2020). Future directions for the concept of salutogenesis: A position article. *Health Promotion International, 35*(2), 187–195.

Bloomfield, M. A., Yusuf, F. N., Srinivasan, R., Kelleher, I., Bell, V., & Pitman, A. (2020). Trauma-informed care for adult survivors of developmental trauma with psychotic and dissociative symptoms: A systematic review of intervention studies. *The Lancet Psychiatry, 7*(5), 449–462.

Chan, S. H. M., Qiu, L., Esposito, G., Mai, K. P., Tam, K. P., & Cui, J. (2023). Nature in virtual reality improves mood and reduces stress: Evidence from young adults and senior citizens. *Virtual Reality, 27*(4), 3285–3300.

Ciangola, M. (2023). The imprint of bodies and threshold in interior architecture: Heterotopias between body and space. *OFFICINA Journal, 41*, 26–33.

Consorti, G., Castagna, C., Tramontano, M., Longobardi, M., Castagna, P., Di Lernia, D., & Lunghi, C. (2023, February). Reconceptualizing somatic dysfunction in the light of a neuroaesthetic enactive paradigm. In *Healthcare* (Vol. 11, No. 4, p. 479). MDPI.

Cotter, K. N., Harrouche, M., Rodriguez-Boerwinkle, R. M., Boerwinkle, M., Silvia, P. J., & Pawelski, J. O. (2023). Virtual art visits: Examining the effects of slow looking on well-being in an online environment. *Psychology of Aesthetics, Creativity, and the Arts.*

Eagleman, D. M., & Perrotta, M. V. (2023). The future of sensory substitution, addition, and expansion via haptic devices. *Frontiers in Human Neuroscience, 16*, 1055546.

England, A., Thompson, J., Dorey, S., Al-Islam, S., Long, M., Maiorino, C., & McEntee, M. F. (2023). A comparison of perceived image quality between computer display monitors and augmented reality smart glasses. *Radiography, 29*(3), 641–646.

Feick, M., Biyikli, C., Gani, K., Wittig, A., Tang, A., & Krüger, A. (2023, October). VoxelHap: A toolkit for constructing proxies providing tactile and kinesthetic haptic feedback in virtual reality. In *Proceedings of the 36th Annual ACM Symposium on User Interface Software and Technology* (pp. 1–13).

Gao, X., Wang, Y., Chen, X., & Gao, S. (2021). Interface, interaction, and intelligence in generalized brain–computer interfaces. *Trends in Cognitive Sciences, 25*(8), 671–684.

Glista, D., Schnittker, J. A., & Brice, S. (2023, June). The modern hearing care landscape: Toward the provision of personalized, dynamic, and adaptive care. In *Seminars in hearing*. Thieme Medical Publishers, Inc.

Grabbe, L. C. (2021). The image becomes a body: Avatarial embodiment in the context of a body ownership illusion. *Virtual Images: Trilogy of Synthetic Realities I, 5*, 218.

Gruber, D. R. (2020). Toward a critical neuroart for a critical neuroscience. *Leonardo, 53*(2), 123–127.

Hutson, J. (2022). The role of virtual reality in art therapy to mitigate autism spectrum disorder (ASD) symptoms. *Open Journal of Clinical & Medical Images, 2*(2).

Karvat, G., Schneider, A., Alyahyay, M., Steenbergen, F., Tangermann, M., & Diester, I. (2020). Real-time detection of neural oscillation bursts allows behaviourally relevant neurofeedback. *Communications Biology, 3*(1), 72.

Kazanskiy, N. L., Khonina, S. N., & Butt, M. A. (2023). Smart contact lenses—A step towards non-invasive continuous eye health monitoring. *Biosensors, 13*(10), 933.

Kourtesis, P., Argelaguet, F., Vizcay, S., Marchal, M., & Pacchierotti, C. (2022). Electrotactile feedback applications for hand and arm interactions: A systematic review, meta-analysis, and future directions. *IEEE Transactions on Haptics.*

Lee, K. K. (2023, November). Affective generative visuals based on data input influenced by user's emotions. In *Proceedings of the International Conference of Innovation in Media and Visual Design (IMDES 2023)* (Vol. 790, p. 387). Springer Nature.

Li, C., & Yip, P. Y. (2023). Remote arts therapy in collaborative virtual environment: A pilot case study. *Frontiers in Virtual Reality, 4*, 1059278.

Liu, W., Zhong, X., Feng, W., Xie, J., Luo, Y., & Guangyuan, L. (2023). *Emotion recognition of virtual reality scenes under olfactory stimulation: An ECG study.* Available at SSRN 4614675.

Lynch, C. R., & Del Casino, V. J. (2021). Smart spaces, information processing, and the question of intelligence. In *Smart spaces and places* (pp. 49–56). Routledge.

Lyon, I., Fisher, P., & Gracey, F. (2021). "Putting a new perspective on life": A qualitative grounded theory of posttraumatic growth following acquired brain injury. *Disability and Rehabilitation, 43*(22), 3225–3233.

Macpherson, F. (2018). *Sensory substitution and augmentation: An introduction.* Oxford University Press.

Magsamen, S., & Ross, I. (2023). *Your brain on art: How the arts transform us.* Random House.

McGorrill, S. (2023). The *benefits of sensory exploration in art therapy with children with social, emotional and behavioral difficulties through a trauma informed lens.*

Mittelmark, M. B., Bauer, G. F., Vaandrager, L., Pelikan, J. M., Sagy, S., Eriksson, M., ... & Meier Magistretti, C. (2022). *The handbook of salutogenesis.* Springer.

Mourtzis, D., Angelopoulos, J., & Panopoulos, N. (2022, March). A teaching factory paradigm for personalized perception of education based on extended reality (XR). In *Proceedings of the 12th Conference on Learning Factories* (CLF 2022).

Pfeifer, P., Hilken, T., Heller, J., Alimamy, S., & Di Palma, R. (2023). More than meets the eye: In-store retail experiences with augmented reality smart glasses. *Computers in Human Behavior, 146*, 107816.

Pilaski, K., & Kaliardos, K. P. (2013). *The Munich Kunstkammer* (Vol. 73). Mohr Siebeck.

Pira, G. L., Aquilini, B., Davoli, A., Grandi, S., & Ruini, C. (2023). The Use of virtual reality interventions to promote positive mental health: Systematic literature review. *JMIR Mental Health, 10*(1), e44998.

Reed, P. G., & Haugan, G. (2021). Self-transcendence: A salutogenic process for well-being. *Health promotion in health care–Vital theories and research* (pp. 103–115).

Ross, A., & Snyder, S. (2023). Weaving it all together: The arts, literacy, adults, and kids. *Childhood Education, 99*(6), 6–13.

Ruívo, M., Frontini, R., & Pernencar, C. (2023). Virtual reality in depressive and anxiety symptomatology-contributions to REVIDA project from a mobile app mapping. *Procedia Computer Science, 219*, 1185–1192.

Schlief, A. M. (2023). *Somatic experience treatment techniques for trauma symptoms: A qualitative case study.*

Shen, V., Rae-Grant, T., Mullenbach, J., Harrison, C., & Shultz, C. (2023, October). Fluid reality: High-resolution, untethered haptic gloves using electroosmotic pump arrays. In *Proceedings of the 36th Annual ACM Symposium on User Interface Software and Technology* (pp. 1–20).

Shvadron, S., Snir, A., Maimon, A., Yizhar, O., Harel, S., Poradosu, K., & Amedi, A. (2023). Shape detection beyond the visual field using a visual-to-auditory sensory augmentation device. *Frontiers in Human Neuroscience, 17*, 1058617.

Shusterman, R., & Veres, B. (2023). Introduction: Somaesthetics and design culture. In *Somaesthetics and design culture* (pp. 1–20). Brill.

Siamptani, M. (2022). *The use of Bio-plastic materials and digital fabrication tools for circular economy products.*

Stepanova, E. R., Desnoyers-Strewart, J., Riecke, B., Huisman, G., & El Ali, A. (2023). *Human bodies as interaction materials for somatic, social, and multisensory virtual reality experiences.*

Tillem, M., & Gün, A. (2023, September). Color blindness in the digital gaming landscape: Addressing critical issues and research gaps. In *European Conference on Games Based Learning* (Vol. 17, No. 1, pp. 817–825).

Tong, Q., Wei, W., Zhang, Y., Xiao, J., & Wang, D. (2023). *Survey on hand-based haptic interaction for virtual reality.* IEEE Transactions on Haptics.

Toya, T., Kobayashi, M., Nakamura, K., & Unoki, M. (2023). Methods for improving word intelligibility of bone-conducted speech by using bone-conduction headphones. *Applied Acoustics, 207*, 109337.

Viviani, G., & Vallesi, A. (2021). EEG-neurofeedback and executive function enhancement in healthy adults: A systematic review. *Psychophysiology, 58*(9), e13874.

Wan, C., Cai, P., Wang, M., Qian, Y., Huang, W., & Chen, X. (2020). Artificial sensory memory. *Advanced Materials, 32*(15), 1902434.

Waisberg, E., Ong, J., Masalkhi, M., Zaman, N., Sarker, P., Lee, A. G., & Tavakkoli, A. (2023). The future of ophthalmology and vision science with the Apple Vision Pro. *Eye*, 1–2.

White, W. F., Burgess, A., Dalgleish, T., Halligan, S., Hiller, R., Oxley, A., Smith, P., & Meiser-Stedman, R. (2022). Prevalence of the dissociative subtype of post-traumatic stress disorder: a systematic review and meta-analysis. *Psychological Medicine*, 1–16.

Wienrich, C., & Latoschik, M. E. (2021). Extended artificial intelligence: New prospects of human-ai interaction research. *Frontiers in Virtual Reality, 2*, 686783.

Williams, T., & Tsien, B. (2013). *Wunderkammer*. Yale University Press.

Zeng, S., Lyu, X., & Kang, J. (2023). Research on innovative method of human-computer collaborative aesthetic education based on hybrid of neuroaesthetics and shape grammar. *IEEE Access*.

Zhong, X., Liu, W., Xie, J., Gu, Y., & Liu, G. (2023). Olfactory-enhanced VR: What's the difference in brain activation compared to traditional VR for emotion induction? *IEEE Transactions on Affective Computing*.

Zuraikat, L. (2020). Google glass: A case study. *Performance Improvement*, *59*(6), 14–20; Ii.

Index

J. Hutson et al., *Sensemaking and Neuroaesthetics*,
https://doi.org/10.1007/978-3-031-58045-1

R

S

The manufacturer's authorised representative in the EU is Springer Nature Customer Service Centre GmbH, Europaplatz 3, 69115 Heidelberg, Germany. If you have any concerns regarding our products, please contact ProductSafety@springernature.com

Printed and bound by CPI Group (UK) Ltd, Croydon, CR0 4YY
01/10/2026
02225894-0006